T0179649

74th Conference
on Glass Problems

74th Conference on Glass Problems

A Collection of Papers Presented at the
74th Conference on Glass Problems
Greater Columbus Convention Center
Columbus, Ohio
October 14–17, 2013

Edited by
S. K. Sundaram

WILEY

Published by John Wiley & Sons, Inc., Hoboken, New Jersey.
Published simultaneously in Canada.

For general information on our other products and services or for technical support, please contact our Customer Care Department within the United States at (800) 762-2974, outside the United States at (317) 572-3993 or fax (317) 572-4002.

Wiley also publishes its books in a variety of electronic formats. Some content that appears in print may not be available in electronic formats. For more information about Wiley products, visit our web site at www.wiley.com.

Library of Congress Cataloging-in-Publication Data is available.

ISBN: 978-1-118-93297-1
ISBN: 978-1-118-93293-3 (special edition)
ISSN: 0196-6219

Printed in the United States of America.

10 9 8 7 6 5 4 3 2 1

Contents

MODELING, SENSING, AND CONTROL

REFRACTORIES I

REFRACTORIES II

Foreword

The 74th Glass Problem Conference is organized by the Kazuo Inamori School of Engineering, Alfred University, Alfred, NY 14802 and The Glass Manufacturing Industry Council, Westerville, OH 43082. The Program Director was S. K. Sundaram, Inamori Professor of Materials Science and Engineering, Kazuo Inamori School of Engineering, Alfred University, Alfred, NY 14802. The Conference Director was Robert Weisenburger Lipetz, Executive Director, Glass Manufacturing Industry Council, Westerville, OH 43082. The themes and chairs of five half-day sessions were as follows:

Batching and Forming
Phil Tucker, Johns Manville, Denver, CO and Ken Bratton, Emhart Glass Research Inc., Windsor, CT

Glass Melting
Glenn Neff, Glass Service, Stuart, FL and Martin Goller, Corning Incorporated, Corning, NY

Modeling, Sensing, and Control
Bruno Purnode, Owens Corning Composite Solutions, Granville, OH and Larry McCloskey, Toledo Engineering Company, Toledo, OH

Refractories I
Matthew Wheeler, RHI US LTD, Batavia, OH and Thomas Dankert, Owens-Illinois, Perrysburg, OH

Warren Curtis, PPG Industries, Pittsburgh, PA and Elmer Sperry, Libbey Glass, Toledo, OH

Refractories II
Andrew Zamurs, Rio Tinto Minerals, Greenwood, CO and Martin Goller, Corning Incorporated, Corning, NY

Preface

In continuing the tradition that dates back to 1934, this volume is a collection of papers presented at the 74th Glass Problems Conference (GPC) published as the 2013 edition of the collected papers. The manuscripts included in this volume are reproduced as furnished by the presenting authors, but were reviewed prior to the presentation and submission by the respective session chairs. These chairs are also the members of the GPC Advisory Board. I appreciate all the assistance and support by the Board members. The American Ceramic Society and myself did minor editing and formatting of these papers. Neither Alfred University nor GMIC is responsible for the statements and opinions expressed in this volume.

As the Program Director of the GPC, I enjoy continuing this tradition of serving the glass industries. I am thankful to all the presenters at the 74th GPC and the authors of these papers. The 74th GPC continues to grow stronger with the support of the teamwork and audience. I appreciate all the support from the members of Advisory Board. Their volunteering sprit, generosity, professionalism, and commitment were critical to the high quality technical program at this Conference. I also appreciate continuing support and leadership from the Conference Director, Mr. Robert Weisenburger Lipetz, Executive Director of GMIC. I look forward to working with the entire team in the future.

S. K. SUNDARAM
Alfred, NY
January 2014

Acknowledgments

It is a great pleasure to acknowledge the dedicated service, advice, and team spirit of the members of the Glass Problems Conference Advisory Board in planning this Conference, inviting key speakers, reviewing technical presentations, chairing technical sessions, and reviewing manuscripts for this publication:

Kenneth Bratton—*Emhart Glass Research Inc. Hartford, CT*
Warren Curtis—*PPG Industries, Inc., Pittsburgh, PA*
Thomas Dankert—*Owens-Illinois, Inc., Perrysburg, OH*
Martin H Goller—*Corning Incorporated, Corning, NY*
Uyi Iyoha—*Praxair Inc.,Tonawanda, NY*
Robert Lipetz—*Glass Manufacturing Industry Council, Westerville, OH*
Laura Lowe—*North American Refractory Company, Pittsburgh, PA*
Larry McCloskey—*Anchor Acquisition, LLC, Lancaster, OH*
Jack Miles—*H.C. Stark, Coldwater, MI*
Glenn Neff—*Glass Service USA, Inc., Stuart, FL*
Bruno Purnode—*Owens Corning Composite Solutions, Granville, OH*
Jans Schep—*Owens-Illinois, Inc., Perrysburg, PA*
Elmer Sperry—*Libbey Glass, Toledo, OH*
Phillip J. Tucker—*Johns Manville, Denver, CO*
James M. Uhlik—*Toledo Engineering Co., Inc., Toledo, OH*
Mathew Wheeler—*RHI US LTD, Batavia, OH*
Andrew Zamurs—*Rio Tinto Minerals, Greenwood, CO*

Batching and Forming

LONG TERM RESULTS OF OXY FUEL FOREHEARTH HEATING TECHNOLOGY FOR E-GLASS FIBERS

Christian Windhoevel[a], Chendhil Periasamy[b], George Todd[b], Justin Wang[b], Bertrand Leroux[c], Youssef Joumani[a]

[a]AIR LIQUIDE Centre de Recherche Claude Delorme
1, Chemin de la Porte des Loges – Les Loges-en-Josas-BP126
F-78354 JOUY-EN-JOSAS Cedex, France

[b]AIR LIQUIDE Delaware Research and Technology Center
200 GBC Drive, Newark, DE USA

[c]AIR LIQUIDE Technology Center (ALTEC)
1, Chemin de la Porte des Loges – Les Loges-en-Josas-BP126
F-78354 JOUY-EN-JOSAS Cedex, France

ABSTRACT

This paper presents the long-term results of ALGLASS ForeHearth (FH) 2-6 kW burner technology in four industrial installations of E-glass fiber and borosilicate container glass industries. ALGLASS FH is an oxy combustion technology developed for glass forehearth that addresses the difficulties encountered in glass forehearth. The ALGLASS FH burner is based on an innovative method for fuel injection with a swirl effect to control flame length (200 to 300 mm). The burner geometry and external body can be easily adapted to customer refractory blocks to meet desired energy profile. Burner robustness, reliability, its flexibility to control flame length and primary energy savings have been confirmed through these references.

INTRODUCTION

Accurate control of molten glass flow temperature in forehearth is crucial to achieve the right conditions, including viscosity at the gob[1]. Air-fuel combustion and electric heating are proven and robust technologies in container and fiber-glass sectors. However, CO_2 and NO_x emission regulations, increase in fuel costs, and production flexibility requirements are pushing glassmakers to investigate new technologies such as oxy-fuel and radiant burners. Oxy-fuel is a mature technology for melting[2]. While literature is abundant on oxygen combustion for melting[3], only a few long-term results on the use of oxy-fuel in glass conditioning during "on the fly" conversion have been published.

Oxygen combustion entails increasing the oxygen concentration in the oxidizer that reacts with the fuel (e.g. natural gas, fuel-oil, coal, synthetic gas). When air is fully replaced by industrial grade oxygen (purity level varying from 80 to 100%, the remaining part being nitrogen and argon), the air burners must be replaced by dedicated oxygen burners. A challenge in glass forehearths is its low available space for retrofitting from air to oxygen combustion. For instance, in a cross-fired furnace, air burners consist of 2 m wide air channels (1.5 m high) with underport installed fuel injectors that let space for changing to oxy fuel burners with a refractory block of 250×250 mm. In forehearths, air

3

burners are installed almost one next to the other. Hence, the oxygen burners' implementation must comply with this available space.

Flame length with oxygen combustion is usually smaller (up to 30% shorter), while the energy released is 3.5 times higher. An optimum oxygen burner must be able to meet such challenges. The burner must also allow for the establishment of a flame without producing a hot spot above the glass flow in order to avoid potential temperature heterogeneity. Further, the flame geometry must be adapted to avoid any flame-to-flame impingement and crown overheating. ALGLASS FH burner has been developed by Air Liquide by taking into account all such specifications[4]. It is compatible with commonly used refractory materials in forehearth such as mullite, AZS and sillimanite.

ALGLASS FH achieves a theoretical thermodynamical fuel saving of 60% when compared to combustion with cold air. In addition, the following benefits have also been observed:

- Better flexibility in energy profile; better control of process
- Lower maintenance
- Safer operation i.e., reduced risk of flashback
- Decreased thermal NO emissions
- Up to 60% reduction in CO_2 emissions

This paper presents the results obtained with ALGLASS FH technology on four industrial installations of existing air-fired and oxy-fired forehearth leg and conditioning zones. As many of fiberglass furnaces are already using oxygen combustion (> 50% of glass furnaces in Europe[5]), operators feel comfortable with its use. For this reason, three of the four references presented in this paper are focused on E-glass. The technology has also been installed in a borosilicate glass forehearth (perfume bottles) with demonstrated benefits on temperature control and fuel savings.

THE ALGLASS FH TECHNOLOGY

The ALGLASS FH burner design is based on a pipe-in-pipe technology that uses a method for injection of fuel with a swirl effect to control flame length. Oxygen injection surrounds the fuel injection (See the design in Figure 1). In order to cover typical needs of glass conditioning processes, the burner power ranges from 2-6 kW (6.8-20.5 MBTU/hr) and the burner external body/geometry can be adapted to fit most customer refractory blocks.

The benefits provided by ALGLASS FH burner were first demonstrated in a pilot furnace. It was shown that flame structure and length can be controlled. The reactants velocities help to control and set the location of the hottest area (or hot spot) of the flame. This parameter is critical to preserve the integrity of both refractory block and burner injector. Selection of suitable swirl effect is important to obtain efficient combustion characteristics. Figure 2 illustrates the principle of the swirl effect on fuel injection. If the mixing between reactants occurs too rapidly (strong swirl), it can lead to shorter flames and localization of the hot spot within the refractory block. This configuration can cause overheating/melting of the block and/or to the degradation of the burner injector due to formation of soot (or carbon) deposits on the burner injector. Sooting gradually changes the flame shape, making the flame shorter and/or less centered within the hole of the block. This can cause flame impingement on the burner block and result in damage to the burner block. On the other hand, when the mixing between fuel and oxygen occurs too slowly (weak swirl), the flame is not robust enough and there is the possibility of inadequate mixing of fuel and oxidizer, which can lead to soot formation on the burner injector.

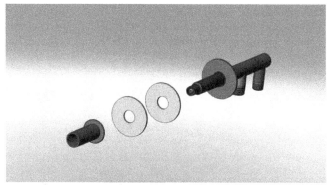

Figure 1: A 3-D scheme representing the ALGLASS FH pipe in pipe burner with fuel (left) and oxygen (right) inlets

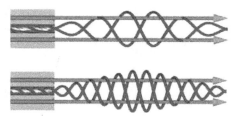

Figure 2: Scheme symbolizing the fuel (red color) and the oxygen (blue color) flows when the power changes; thanks to the swirl effect, the reaction between reactants takes place in the same area which leads to the same flame length

The swirl effect helps to maintain the hot spot of the flame at the same location even when the burner power is varied within a wide range. For example, when the power increases, the swirl effect on the fuel injection also increases which, in turn, increases mixing between the reactants, thus avoiding change in the flame length and the location of hot spot. Figure 2 illustrates the principle of the swirl effect on fuel injection.

Another benefit of this ALGLASS FH design is its capability to improve and optimize heat transfer to the glass. The flame hot spot is localized outside the refractory block by design, but not too far from the block outlet. Consequently, the maximum heat is transferred to the glass close to the forehearth walls to compensate for heat losses through the walls of the forehearth line and to improve the homogeneity of the glass.

Pilot furnace trials have demonstrated the following advantages:
- Compatibility with different refractory blocks materials: AZS, mullite and sillimanite burner blocks have been tested and showed no issues when operated with burner power < 1.5 times nominal.

- Adaption to different customer block geometry: in pilot furnace atmosphere, no perturbations of flames or soot deposits were observed.
- Zero maintenance: in a clean environment, such as a pilot furnace, the burners have been tested during several weeks at 1400°C (block temperature) without observable damage.
- Eliminate safety risks due to reactants premixing: thanks to separate injection of gas and oxygen, flashback inside natural gas line cannot occur. Moreover, the swirl allows a perfect start-up even at low power avoiding flame lift-off.

PREPARATION FOR INDUSTRIAL TRIALS

In order to prepare for each industrial trial, the customer's burner block (corresponding to the specific forehearth line) was first installed and tested with ALGLASS FH burners in the a pilot furnace without glass. The following thermal diagnostic equipment was used:

- Thermocouples – to measure the longitudinal temperature profile at the top of the burner block. This helps to determine the position of the hot spot.
- Pyrometer – to measure radial temperature profile of the refractory block around the outlet hole. This allows to assess the homogeneity of the heat transferred all around the block and also to verify that there is no damage to the front face of the block.
- Temperature sensors - installed at the chamber bottom to simulate industrial glass surface temperature. This helps to understand the temperature profile, hence heat transfer to load.
- Pressure sensors – for both natural gas and oxygen inlets.

From the pilot furnace tests, the optimal range and recommended burner power were identified, resulting in improved heat transfer to the glass while minimizing the risk of damage to both refractory block and burner injector.

In addition to these experimental studies, numerical simulations were also made with in-house computational fluid dynamics (CFD) software called ATHENA[6]. This tool was used for two main purposes: to test the effect of burner parameters on fluid flow and radiation that are normally difficult to measure experimentally and to simulate full forehearth operation and study the thermal interaction of the burner with forehearth. Turbulence, combustion and radiation models included in ATHENA software have been validated for swirled oxy-combustion, through comparison of measured and calculated data.

INDUSTRIAL RESULTS

Reference 1: Burner trials in leg zone of fiberglass forehearth

The ALGLASS FH technology was implemented in the conditioning leg of an E-glass fiber manufacturer in Europe; previously, the conditioning leg was operated with premixed air fuel burners. The oxy-fuel burner was operated in a range of 5 kW (17 MBTU/hr). During this trial several issues were observed:

- Variations in the customer refractory block opening for burner. As a mitigation strategy, the ALGLAS FH burner was equipped with "auto centering" modules to have the burner straight even if the opening is not straight.

- Burner material issues possibly related to high metallic body temperatures. Three burners were equipped with three thermocouples each and the temperatures were recorded. The temperatures were in the expected range with the burner tip being hottest. To reduce any possible risk of overheating, the burner body material was changed to a high heat resistance material.
- Temperature increase at one burner. Upon inspection, it was found out that the high-density energy release by the flame inside the long and narrow refractory block channel could not be dissipated to the surrounding refractory and has caused the issues.

The geometry of the prevailing burner refractory block at the customer site is critical for the success of the ALGLASS FH implementation in a refining zone or conditioning leg. During the operation, the easy installation of the burner was positively noted as well as no adverse effects on glass quality.

Reference 2: *Full conversion of a conditioning zone of fiberglass forehearth*

In 2006, Air Liquide started a partnership with one of its customers to apply ALGLASS FH technology in one of its forehearth lines. Implementation of the technology initially focused on the conditioning zone. The trial objectives were: (i) to evaluate the firing flexibility of ALGLASS FH burner to forehearth demand; (ii) to check the robustness of the technology during several months' operations, and (iii) to quantify the impact of the technology on the thermal behavior of the forehearth.

In order to prepare for the trials in the chosen zone of the conditioning channel, where the temperature of the glass is close to 1,250°C (2,282°F), a representative refractory block from customer was installed and tested in the Air Liquide pilot furnace for several hundreds of hours. The burner was tested in the range 2-6 kW in order to evaluate the differences in temperature profiles (Figures 3 and 4). Because the burner must fit within the customer's burner block, the burner inlet required some minor modification, in order to fit with the block geometry. The trials also provided an opportunity to verify the installation procedure.

Figure 3 shows the temperature profiles measured on the block top for three different powers. Although the power of the ALGLASS FH burner increases, the maximum temperature remains below 1,300°C (2,372°F) and the profile remains similar. Figure 4 presents the temperature profiles measured on the bottom of the chamber of the pilot furnace. The temperature close to the block outlet also remains close to 1,250°C (2,282°F) - demonstrating that the location of the hot spot does not change when power varies from nominal to 1.5 times nominal. When the power is increased, a more homogeneous temperature profile is obtained up to a distance representing the middle of the channel.

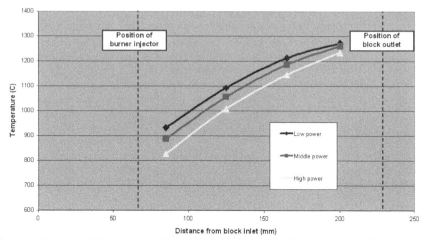

Figure 3: Graphic – Tests in the AIR LIQUIDE pilot tests – Temperature profile on block top when the ALGLASS FH burner was operated at three different powers inside the customer's refractory block – for E glass fiber process

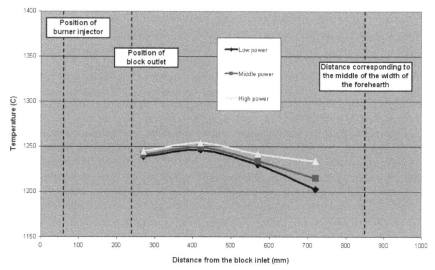

Figure 4: Graphic – Tests in the AIR LIQUIDE pilot tests – Temperature profile on chamber bottom when the ALGLASS FH burner was operated at three different powers inside the customer's refractory block – for E glass fiber process

The industrial trial period with 12 ALGLASS FH burners took 18 months. ALGLASS FH burners were implemented in a conditioning channel. The burners were ignited and used with an oxygen/natural gas ratio very close to stoichiometric conditions. The flame was checked through peepholes; it remained well centered inside the block. In the beginning, there had been minor random deposits looking like glass noted in the lower side of the burner body. However, no further deposition issues were notes after the initial start up.

Figure 5 shows pictures of one ALGLASS FH burner: on the left side, the burner is operated inside the forehearth line, the well centered flame inside the refractory block can be noticed; on the right side, the burner was pulled out for inspection after several weeks of continuous operation; there was no sign of burner damage or glowing injector.

Figure 5: Pictures of one ALGLASS burner: operated inside the E glass fiber forehearth line (left side) and pulled out for inspection after several weeks of operation (right side)

During the entire 18-month period of operation, the flame shape remained consistent and there no damage to the burner blocks was noted. Furthermore, due to pull rate variation, the burner power was changed by 50% of its nominal rate. The temperature of glass and channel were checked and the energy profile easily adapted to overcome potential overheating. Today, the forehearth is operated with around 100 ALGLASS FH burners since April 2010.

Reference 3: Full conversion of a leg of fiberglass forehearth
The E-glass producing customer was interested in assessing the performance of the ALGLASS FH technology - with regard to robustness and fuel savings. The customer ran trials of the technology first in the refining zone and then later in the two legs of the conditioning zone and fuel savings. For the refining zone, 8 ALGLASS FH burners were trialed for one month, starting in October 2010.

As with the earlier trials, in order to prepare the trials in the refining zone, where the temperature of the glass is close to 1,285°C (2,346°F), a representative refractory block from customer was installed and tested in the AIR LIQUIDE pilot furnace. The required power was known; therefore, the burner was tested within this range to appreciate the differences in terms of temperature profiles. See Figures 6 and 7 for pilot furnace results.

Figure 6 shows the temperature profiles measured on the block top for three different powers. Although the power of the ALGLASS FH burner increases, the maximum temperature always remains below 1,304°C (2,380°F) and the profile remains similar. Figure 7 presents the profiles of temperature measured on the bottom of the chamber of the pilot furnace. The temperature close to the block outlet remains close to 1,298°C (2,370°F) - demonstrating that the location of the hot spot remains fixed.

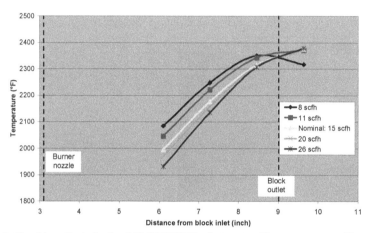

Figure 6: Graphic – Tests in the AIR LIQUIDE pilot tests – Temperature profile on block top when the ALGLASS FH burner was operated at five different powers inside the customer's refractory block – for E glass fiber process

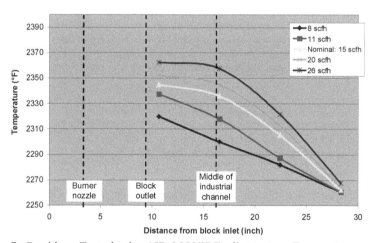

Figure 7: Graphic – Tests in the AIR LIQUIDE pilot tests – Temperature profile on chamber bottom when the ALGLASS FH burner was operated at five different powers inside the customer's refractory block – for E glass fiber process

Industrial operation of around 50 ALGLASS FH burners in the entire forehearth leg began in November 2010. ALGLASS FH burners were implemented in a 609 mm to 304 mm (24" to 12") roughly wide zone of the conditioning channel. The burners were ignited and used with an oxygen/natural gas ratio very close to stoichiometric conditions. After a short troubleshooting phase (clogging and deposits), all the burners have been run successfully, without further issues.

As with the previous references, the trials showed no adverse effect on burner refractory block or glass quality (e.g. reboil). It was also noted, that the flame length was constant over varying natural gas flow rates, ensuring that the hot spot remained outside of the refractory block. Robustness of the burner was assessed through regular visual inspections, which showed no deposits and clean burner bodies and tips.

Reference 4: *Conversion of one zone in a borosilicate glass forehearth*

A glass manufacturer specializing in the production of borosilicate container glass for the cosmetic industry wanted to improve the quality of the glass while having the ability to control the very low thickness of certain ranges of bottles. When using air combustion, reboil issues in the forehearth line occurred, particularly in the zone downstream of melter exit, where the glass temperature is close to 1,300°C (2,372°F).

In preparation of the oxy-conversion, CFD calculations were performed using ATHENA software to choose the best conditions for the trials: The main results of simulations relating to the ALGLASS FH technology showed that a global power close to 50-53 kW (170-180 MBTU/hr) was suitable for oxy-combustion.

Figure 8 shows the interactions of heat transfer between two face-to-face flames, which occur in the case of air burners (top side of the figure); generating a temperature close to 1,700°C (3,092 F) near the crown of the forehearth. Using ALGLASS FH burners, (bottom side of the figure) the temperature near the crown would remain close to an acceptable level of 1,500°C (2,732°F). Figure 9 presents one part of the zone downstream of the melter exit; the impingement between flames occurs with air burners (left side of the figure) and the temperature of the glass surface does not appear uniform along the width of the chamber. The temperature reaches 1,600°C (2,912 F) in the middle of the channel, which represents a significant risk of glass reboil.

Using ALGLASS FH burners (right side figure 9) the temperature of the glass surface appears homogeneous in the area close to the channel walls, whereas the maximum temperature reached in the middle zone is not higher than 1,400°C (2,552°F), preventing the glass reboiling phenomenon.

In the AIR LIQUIDE pilot furnace, the customer block was operated and temperature profile of the block indicates that the hottest level is around 1,325°C (2,417°F) and positioned outside of the block. The temperature profile on the chamber bottom of the pilot furnace shows that the temperature remains quite uniform and very close to 1,300°C (2,372°F) near the block outlet and into the half-width of the industrial forehearth line; indeed, the energy is homogenously transferred towards the bottom of the chamber. The temperature of the block, around the outlet hole, is only 20-25°C higher than the chamber bottom.

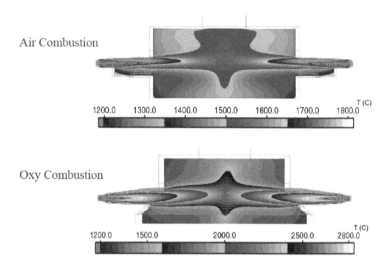

Figure 8: Modeling of two face-to-face burners in the zone downstream of the melter exit – for borosilicate container glass process: air combustion case (top) and oxy combustion case (bottom)

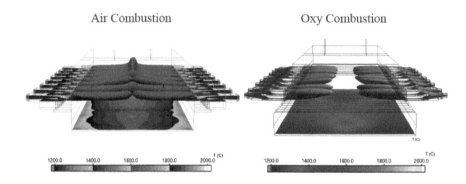

Figure 9: Modeling of face-to-face burners in one part of the zone downstream of the melter exit – for borosilicate container glass process: air combustion case (left side) and oxy combustion case (right side)

The industrial trial period in the refining zone with 8 ALGLASS FH burners started in August 2008. ALGLASS FH burners were installed in face-to-face configuration in the 400 mm (15.75") approximately wide zone downstream of the melter exit the oxy-combustion zone is positioned close to the air combustion zones (Figure 10). The objective was to bring energy to the glass exiting the melter to improve the homogeneity of the glass while avoiding the reboiling of glass.

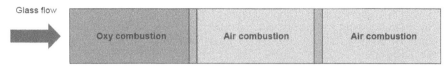

Figure 10: Schematic top view of the first part of the borosilicate container glass forehearth line where oxy combustion zone, downstream of the melter exit, is represented (left side)

The oxygen/propane ratio was higher than stoichiometric conditions to keep the same conditions as in the air combustion zones and to satisfy the customer's request to maintain an oxidizing atmosphere inside the forehearth. In contrast to the air combustion case, there was no reboil of glass using the oxy-combustion. Moreover this conversion had a significant improvement of the quality and of the homogeneity of the glass at the gob; the control of small thicknesses was improved, what allowed decreasing the rate of rejects. The ALGLASS FH burners have been using in this zone of the forehearth for more than five years. Figure 11 presents the picture of one ALGLASS FH burner operated outside the forehearth line during an inspection.

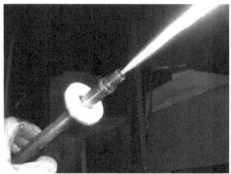

Figure 11: Picture of one ALGLASS FH burner operated outside the borosilicate container glass forehearth line

CONCLUSIONS

During the last 6 years, Air Liquide has performed four long-term trials and commercial references with using oxygen in glass forehearths. The following conclusions are drawn for the different types of glass:

For the E glass fiber process: ALGLASS FH burners in one zone of the conditioning channel replaced several oxygen burners. The flames shape remains good; there has been no damage to the burner blocks.

For the borosilicate container glass process: The ALGLASS FH burners using natural gas or propane and oxygen have been installed face-to-face replacing air burners and using the same refractory blocks. In contrast to air combustion case, there was no reboiling of glass. Moreover this conversion significantly improved the quality and the homogeneity of the glass at the gob, improved the control of small thickness bottles, and reduced the amount of rejected material. ALGLASS FH burners are still in operation today. After more than 6 years of operation, the main benefits of ALGLASS FH technology are:

- Primary energy reduction by up to 60%
- Decrease in CO_2 emissions by up to 60%
- Better process control and homogeneity of the glass
- Better heat transfer to the glass without negative impact on quality (e.g. no reboil)
- Easy and simple installation also in existing forehearth, thanks to existing suitable burner block geometry database
- Robustness and reliability

REFERENCES

1. Glass Industry in the future, Office of industrial technologies, April 2002
2. K. Cook, Forehearth profit from oxy-fuel combustion, Glass International, September 2007
3. C. E. Baukal, Oxygen-Enhanced Combustion, Second Edition, CRC press, March 15, 2013
4. R. Kalcevic, R. Tsiava, R. Fujinuma, C. Periasamy, R. Prabhakar, G. Todd, Industrial results of ALGLASS FH oxy fuel ForeHearth burner operation, Glass Problems Conference, 2008
5. European Commission, Glass Manufacturing Industry Integrated Pollution Prevention and control, July 2009
6. Till M., Marin O., Louédin O., Labégorre B., "Numerical Simulation of Industrial Processes", American/Japanese Flame Research Committees, International Symposium, Kauai, Hawaii, September 9-12, 2001

GLASS PRODUCTION LOSSES ORIGINATING FROM CONTAMINANTS IN CULLET AND RAW MATERIALS

J. Terry Fisk
JTF Microscopy Services, LLC
Hammondsport, NY USA 14840

ABSTRACT
Not all solid inclusions in glass products are due to incomplete melting of batch ingredients, refractory erosion, corrosion or spalling, air surface volatilization from the melt, or "cold glass" devitrification. Raw materials may contain parts-per-million levels of "refractory" mineral grains and there may be unwanted materials present in the cullet, especially if externally sourced cullet is used. Comprehensive analysis of the inclusions by petrographic methods using transmitted polarized light microscopy (PLM) and/or sometimes augmented by qualitative elemental analysis via scanning electron microscopy with energy dispersive x-ray spectroscopy (SEM/EDX) can reveal the true root cause of the problem. This paper will present several examples illustrating the analytical techniques and some of the revelations that can be provided by those analyses.

INTRODUCTION
It is a well-recognized fact that it is impractical if not totally impossible to make glass that is entirely free of any inhomogeneity defect. There is usually some low frequency of a combination three basic category of defects in the glass product such as: gaseous defect (seed, blister and airline), vitreous defect (cord, ream and knot) and solid inclusion (crystalline stone and metal/oxide inclusion). Each of these needs to be appraised within the constraints of the quality specifications for the given product. Some combination of engineering practices such as process control charting and laboratory evaluation of glass defects should be done on a routine basis to monitor the glass-making process and provide a baseline of the "normal" operating status quo with respect to the aforementioned glass defects so that when a quality "upset" does occur it can be understood and remedied as quickly as possible.

Once a working baseline has been established for the usual solid inclusions, the laboratory technologist can more readily recognize the unusual stones and provide that information to the batching and melting personnel. Batching and melting departments can then focus their efforts on finding the cause of the change that led to the upset. This in turn may point out one or two suspect raw material or cullet shipments that may need to be fully evaluated in the laboratory to deliver the proof that pinpoints the actual root cause or source of the upset. The purpose of this paper is to describe some investigative tools and processes that work to help those who are dealing with production upsets caused by the crystalline stone category, especially ones originating from raw material and cullet contaminations.

STONE DEFECTS PROCESS: ANALYSIS, SOURCING AND VERIFICATION
Base Case Typified by the "Usual" Stone Categories
The usual background level of 1 – 10% glass loss for stones can include quite a wide variety of substances from an equally wide variety of source/mechanisms. It is nearly impossible to provide a thorough list of the possible background stones here because of the many permutations with regards to the assorted glass chemistries, furnace designs, refractory material utilizations, and forming methods used throughout the world of glass production. The following list is a generic one that does not get into the specifics:
• Incomplete melting of batch materials – especially silica batch stones

- Glass refractory interactions – corrosion interface layers and alteration products
- Refractory erosion – surfaces, joints and exposed corners
- Refractory spalling – usually due to thermal shock or physical constraint
- Air surface volatilization – B2O3 and alkali loss resulting in "silica scum" stones
- "Cold glass" devitrification – crystallization of a glass due to sub-liquidus temperatures
- Process related contaminants – metal/oxide inclusions from batch handling equipment, screw feeders, burner tips, electrodes, thermocouples, stirrers, etc.

"Upset" Condition Typified by the Presence of "Unusual" Stone Categories

Something changes and the normal stone loss of 1 – 10% has increased to a higher level. It may be that the upset is due to any one or more of the aforementioned stone types because furnace operational parameters (temperatures, fill/pull rate, etc.) have changed. Or it may be that something completely different has occurred. Collect samples, preferably more than ten samples, and get them to the laboratory technologist without delay. Petrographic analysis should be able to recognize the usual suite of stone defects and identify any new or unusual stones, if present.

As was noted above, it would not be possible to list all of the possible "unusual" stone defects that may arise from raw material and cullet contaminant materials. A basic list of source conditions is furnished below while a more detailed list can be found in later sections of this paper.

- Raw material contaminants: concentrations as low as a few ppb, up to whole percentages
  - Normally occurring "tramp" mineral grains – may become a problem if the frequency or size of these in the bulk raw material is elevated
  - Unwanted "refractory" mineral grains – certain naturally occurring minerals such as Chromite, Corundum, etc.
- Cullet contaminants: concentrations usually in the ppm to tenths of a percent range
  - Unwanted foreign materials – non-glass substances, recirculated refractory stones
  - External/foreign cullet can be more troublesome
  - Some of these compounds may be able to survive multiple trips through the melter by nature of their size and refractoriness (i.e.: their resistance to attack)

Stone Identification Process

Considering the fact that the process is invariably contributing some number of "usual" background stones even during an upset, melting/forming personnel will need to ensure that the laboratory has a good selection of stone specimens to evaluate. Any analysis of stone defects, whether it be for baseline monitoring or for a quality upset investigation, should be started with an absolute minimum of twenty specimens, and the case could be made that fifty specimens might be a better target number. Be sure to include samples that encompass the full size range and possible color range of stones that are being rejected. This will ensure that the one or two types of stones that truly comprise the upset will be captured and included in the evaluation.

Petrographic analysis should begin with a low magnification examination of each stone in-situ (in-glass) to describe the degree of angularity or rounding of the crystalline matter, natural color, the size of the solution sac, and presence or absence of outgassing from reaction of the glass upon the foreign particle. However, the main purpose of this initial microscopic examination is to sort the samples into their respective categories based on the dominant crystal phase(s) observed. Only the exterior/reaction layer of the overly large and/or optically dense specimens can be evaluated by this method. The main benefit of this initial examination is that it gives a quick indication as to what the major defect type(s) are and which areas the technologist should focus his or her detailed analysis efforts on. Thin-sections should then be prepared of several stone samples from each of the major categories previously separated so that a detailed petrographic analysis can be conducted on the materials present within the interior of the stones. The detailed petrographic analysis should include crystal phase identification based on observed optical properties and a morphological differentiation of primary (natural) and secondary (recrystallized) forms of those phases. Once the crystal phases have been identified the technologist will have a reasonably good understanding of the chemical composition of the stones as well as some explanation as to why certain secondary phases were produced by reaction of the glass chemistry with that of the original defect substance. This should also lead to conclusions about the source of the defects analyzed.

In the event that petrographic analysis cannot completely identify the components of the stone defects then the top surface of the thin-sectioned specimens can be fine-ground and polished such that elemental analysis can then be done by SEM/EDX or Microprobe. Irrespective of the cost of such electron microscopes with associated x-ray spectroscopic systems and having an in-house operator with the knowledge to properly do so, the time involved in sample preparation and analysis can be up to five times lengthier than that required for complete petrographic analysis.

Once the analysis is done and the information communicated back to melting/forming then some evaluations of the process with respect to recent changes must be undertaken. For example, if the major stones comprising the upset were from batch or refractory sources then engineers should look for changes in furnace operating temperatures, hot spot movement, changes in fill or pull rate, minimum residence time, etc. But if the upset was due to "unusual" stones then the questions should be geared toward evaluating the timing of the upset and arrival of new shipments and incorporation of raw materials or cullet. Was there a supplier change? Did the supplier make a change in his process? Is there anything out on line with regards to chemistry or size distribution of those materials?

Table 1 is a listing of numerous raw material and cullet contaminants that the author has encountered during a lifetime of work in this area. Chemistries often vary from ideal formulae

Table 1. Raw materials vs. cullet contamination

Raw Material Contaminants	Cullet Contaminants
Chromite (FeO-Cr_2O_3 w/Mg, Al, Ti, etc.)	Clay/Ceramic (Coffee mugs, pottery)
Zircon (ZrO_2-SiO_2)	Pyroceram (High temperature glass-ceramic)
Corundum (α-Al_2O_3)	Carborundum (Silicon carbide abrasives)
Garnet ($3CaO$-Al_2O_3-$3SiO_2$ w/ Fe, Mg, Cr, etc.)	Corundum (Alumina abrasives, refractory raw materials, sand blast materials, etc.)
Spinel (MgO-Al_2O_3 w/Fe, Zn, Ti, Cr, etc.)	Metals – Aluminum alloys and any of the non-ferromagnetic stainless steel alloys
Andalusite (Al_2O_3-SiO_2)	Chromite stones in the glass
Kyanite (Al_2O_3-SiO_2)	Fuse cast AZS refractory stones
Sillimanite (Al_2O_3-SiO_2)	Bonded AZS refractory stones
Feldspar (($Na,K)_2O$-Al_2O_3-$6SiO_2$)	Alumina refractory stones
Quartz (SiO_2)	Zircon refractory stones
Xenotime (YPO_4)	Tin oxide refractory stones

Confirmation of Raw Material or Cullet Contamination

The above analyses and process evaluations may indicate that the most likely cause for the current glass defect quality issue has been delivered, as it were, concealed within a raw material or perhaps a cullet shipment. Now it is time for the laboratory technologist to go back to work to hopefully confirm or deny the possible source of the upset.

Depending on the raw material being investigated there are different ways to go about the discovery process. If the problem is believed to be chromite in sand then the plan of attack may be to manually sift through a large petri dish full of sand while peering through a low powered stereobinocular microscope, plucking out the dark black colored grains with a fine pointed set of tweezers. If the stones from the upset are larger than 500 micrometers in size and it seems the sand may be the source then screening the sand through a US 30 Mesh sieve will eliminate the "fines" and improve one's chances of satisfactorily isolating the contaminant material. A strong rare earth magnet can be used to separate weakly ferromagnetic minerals such as Chromite and to lesser degree alumina abrasive grit particles. Hot water dissolution can be leveraged in the case of certain raw materials such as soda ash that are soluble, and the contaminant particles can be collected on a filter paper. Hydrochloric acid works to dissolve limestone and dolomite raw materials leaving behind an assortment of insoluble mineral grains that were concomitantly deposited by the sedimentary process that later became the host rock. That assemblage of insoluble substances can then be screened through a US 30 Mesh sieve to further concentrate the +600 μm fraction that would be much more likely to survive a trip through the glass furnace system. Since most of the "refractory" minerals are much denser than the raw materials commonly used in glass batch mixtures, a heavy mineral separation (HMS) can be performed to sort out the particles to be analyzed. Furthermore, depending on the size of the stone defects being encountered in the upset, a suitable screen can be used on the raw material to eliminate either the fine fraction or coarse fraction of that raw material before the HMS. Additionally, if the HMS produces a large amount of "heavy" mineral grains then these may be screened after the HMS to further concentrate the particle size range that makes the most sense.

Regardless of what technique was used to concentrate the portion of the raw material that is under investigation, the next step of the analysis is to perform petrographic analysis of the grains by transmitted PLM. This can be done by transferring some of the grains onto a microscope slide, adding a few drops of refractive index liquid of known index, and sealing with a coverslip. SEM/EDX analysis can be performed on another portion of the assorted grains, or if in limited supply, after washing and drying of the grains previously analyzed by PLM.

If the suspected source is in the cullet the process is slightly modified because of the size range of the glass cullet itself. Manually sorting out oddities and abnormally colored pieces may help to isolate foreign materials such as pottery, Pyroceram, grinding discs, tools, soda cans, etc. The author has resorted to using the "panning for gold" technique assisted by water and an upturned Frisbee to isolate smaller and denser foreign materials such as Carborundum, Corundum, metals, etc. Unfortunately, it can be very time consuming and difficult to find and isolate refractory stones that may be encased within the glass pieces as a result of the original manufacturer's glass quality upset having been discarded and turned in for recycling. In this event there may be some high level discussions with the company or consortium that supplies the cullet.

PLM analysis will need to be performed on any suspicious particles that have been isolated from the cullet samples. Larger chunks and any stones that might have been found encased within the glass pieces will likely require thin-sectioning to facilitate the analysis. Smaller grains may be examined after having been transferred to a microscope slide and prepared as described above. SEM/EDX analysis may also prove to be advantageous for some rather obvious reasons.

Whatever the result, the final part of the investigation is to ask the thousand-dollar question, "Does the size and chemistry of what has been found in this raw material (or cullet) make sense with respect to that of the stone defects being experienced?"

EXAMPLE CASE HISTORIES
Chromite in Fluorspar Raw Material
This upset was typified by black stones ranging from 0.2 to 1 millimeter in size accounting for up to 15% loss in a white opal dinnerware product. The larger stones were causing breakage of the ware in the lehr. The stones were found to consist of secondary Cr_2O_3

crystallization surrounding a primary core of chromite – a dark red-brown isotropic substance. There were no chrome-bearing refractories used in the furnace construction. Chromite grains were found in the Fluorspar (CaF2) raw material by visual sorting using a stereobinocular microscope and fine pointed tweezers. Chromite was estimated to be present at about 100 ppm in the Fluorspar even though chromium was within spec based on chemical analysis of the bulk raw material. The supplier later admitted that the raw material had been contaminated at the shipyard by the use of "dirty" handling equipment used previously on a shipment of refractory grade Chromite.

An example PLM photomicrograph, SEM micrograph and EDX spectrograph of one of the Chromite stone inclusions are included as Figures 1 through 3 below. Figure 4 contains an SEM micrograph of one of the Chromite grains that had been separated out of the Fluorspar raw material sample.

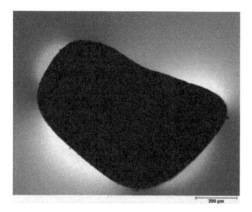

Figure 1: Chromite Inclusion in glass; original magnification ~ 150X

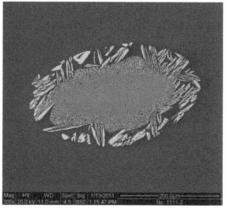

Figure 2: SEM micrograph of Chromite inclusion in glass, cross-sectioned and polished; original magnification 300X

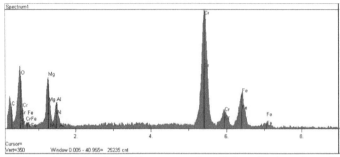

Figure 3: EDX spectrograph collected from core region of Chromite stone shown in Figure 2 (carbon coated)

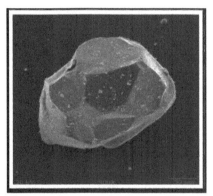

Figure 4: Chromite grain after separation from the Fluorspar raw material; original magnification 300X

Spinel in Dolomite (limestone)

A sudden onset of markedly increased stone losses occurred in a soda lime float operation. The stones ranged from 0.5 to 2.0 millimeters in size. Petrographic examination of the stones found them to be well-rounded single grains of a pale green, isotropic substance surrounded by a Nepheline (Na_2O-Al_2O_3-$2SiO_2$) fringe and encased in large solution sacs. The conclusion was that the primary core material was Spinel (MgO-Al_2O_3) with suspected solid solution addition of iron oxide based on the pale green color of the isotropic mineral when viewed in plane-polarized light. The composition was confirmed by SEM/EDX analysis. The customer suspected the source of the upset to be the recent lot/shipment of Dolomite and included a two-pound bag of that material along with the stones. Spinel was positively identified in the hydrochloric acid insoluble fraction obtained from the Dolomite sample. Spinel is known to be present as a tramp level accessory mineral within metamorphic Dolomite and Limestone deposits, sometimes concentrated in small veins within the rock. The raw material supplier had recently changed to a new location on the opposite side of the pit mine. Bulk chemical analysis of the raw material showed only a slight uptick in alumina concentrations, but still well below the agreed limit established in the glass manufacturer's specifications. The supplier switched to another mine site to avoid the spinel in future shipments.

Figures 5 and 6 contain a PLM view of one of the Spinel stones in glass and an SEM micrograph of the Spinel that was sorted out of the Dolomite raw material after hydrochloric acid dissolution and heavy minerals separation of the resultant insoluble residue.

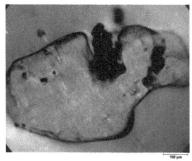

Figure 5: Spinel inclusion in soda lime float glass; original magnification ~ 175X

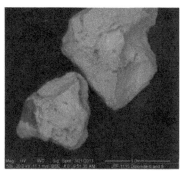

Figure 6: SEM micrograph of Spinel grains from HCl acid insoluble fraction of Dolomite; original magnification 50X

Clay-Ceramic Particles in Cullet

A soda lime container manufacturer experienced a stone upset and breakage of the bottles shortly after forming. The upset suddenly arrived during the night shift and persisted for five days with seemingly no improvement after several furnace moves made during that timeframe. The customer had assumed the stones were un-melted silica batch material and the furnace moves were aimed at improving that condition. Petrographic analysis quickly disproved the silica batch stone theory and identified that the stones were actually a low grade alumino-silicate clay-ceramic product such as that found in coffee mugs and pottery. The highly-stressed stones measured up to 8 millimeters in diameter and had large solution sacs with numerous out-gassing bubbles and large Nepheline borders surrounding a nearly amorphous clay-like core material. There was also a fair amount of tiny quartz grain additive dispersed throughout the clay-like phase. A manual sorting of the cullet resulted in the finding of numerous chunks of opaque white and tan colored material that should not have been there. Petrographic analysis confirmed these were similar low-grade clay-ceramic material as had been identified at the interiors of the stones.

The root cause of the contamination was coffee mug fragments that had been crushed along with the host glass in a post-consumer recycled "glass" supply.

The public does not have any idea what havoc they can inflict upon glass manufacturers when coffee mugs, pottery, glass-ceramic articles, non-soda lime glass products and metallic bits are lackadaisically discarded into the so-called "glass" recycling bins.

PLM photomicrographs of two of the thin-sectioned clay-ceramic stones with large Nepheline alteration fringes are included in Figure 7 below. A high degree of out-gassing bubbles can be seen at the inner portion of the Nepheline fringe in the example at the left hand side of the figure whereas there appears to be only one out-gassing bubble captured in the plane of the thin-section for the second example to the right. A photomicrograph of one of the thin-sectioned clay-ceramic chunks is included in Figure 8.

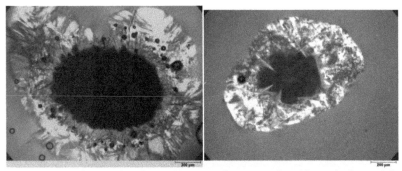

Figure 7: PLM photomicrographs of two Clay-ceramic stones after thin-sectioning; original magnification ~ 50X

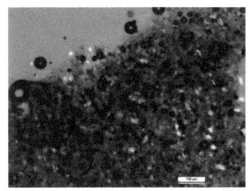

Figure 8: Pottery shard isolated from cullet supply; thin-sectioned; original magnification ~100X

Corundum Alumina in Cullet

Two soda lime container manufacturers, one located in New Jersey and the other in Pennsylvania, experienced stone upsets at nearly the same time and samples were delivered from each company on the same delivery truck. The problem at both factories was Corundum alumina stones that were occurring as coarse single grains of Corundum with a blue-brown coloration when viewed in plane polarized light. All indications were that the original substance comprising the stones was an alumina abrasive grain or refractory grade alumina product. Each case was handled discretely but both companies came to the same conclusion that their new cullet supply was the source of the problem.

An independent consultant engineering firm was brought in to help with the investigation and provided samples of the cullet about two weeks later for analysis. Samples had been taken from several discrete locations in and around the mountain of cullet. Knowing that we were looking for some form of alumina, the first step of the investigation involved a separation of the alumina grains from the cullet by a water-assisted "panning for gold" technique. Fused brown alumina was found to be fairly evenly distributed around the surface of the cullet pile but the contamination rapidly trailed off just a few inches beneath the exterior surface of the pile. The ultimate cause of the contamination was deemed to be poor attention to detail at the shipyard and storage facility. The consortium that owned the cullet had left explicit instructions as to how the cleanliness of the cullet was to be maintained. Unfortunately, the shipyard personnel ignored those instructions and had placed the glass cullet on a holding pad that was adjacent to and downwind to a similarly large pile of fused brown alumina. The ensuing lawsuit was settled out of court shortly after expert witness depositions. The cullet supplier plaintiff was able to recoup losses and maintain a good relationship with the glass manufacturers involved.

Figure 9 contains a PLM photomicrograph of one of the Corundum alumina stones in a soda lime container product. Figure 10 contains a pair of PLM photomicrographs of some of the fused brown alumina grains that were separated out of the cullet samples.

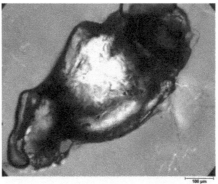

Figure 9: Corundum alumina stone in soda lime container glass; thin-sectioned; original magnification ~100X

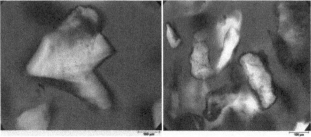

Figure 10: Fused brown alumina grains from cullet samples; original magnification ~100X

CONCLUSIONS

The persons involved in the glass making operation must first be able to recognize when there is a glass defect upset taking place. Process control charts and baseline analyses of normal production defects should be available for review and consulted during the upset. Samples from the upset should be collected and sent to the laboratory for analysis. Those results should be compared to normal baseline conditions so that the variety(s), frequency and size range of defect(s) responsible for the quality upset can be sorted out and efforts focused on the elimination of the problem(s). Any recent changes made in operational set-ups and/or materials utilization should absolutely be reviewed. With any luck, the laboratory analyses, production status reviews and timeline of raw materials and cullet shipments may point to one or more of the high probability sources that should be investigated further. The short list of suspected sources can then be explored by inquiry with suppliers while additional analysis are being completed on any and all materials that are available. Regardless of the outcome, the results should be thoroughly documented to provide a historical database in the event of a recurrence of the problem as well as to provide a basis of understanding when negotiations are being made with the suppliers.

REFERENCES

1. Clark-Monks, C. and Parker, J.M., Stones and Cord in Glass, Society of Glass Technology, Sheffield, 1980

2. Begley, E.R., Guide to Refractory and Glass Reactions, Cahners Publishing, Boston, 1970

3. Ford, W.E and Dana, E.S., A Textbook of Mineralogy, John Wiley and Sons, New York, 1954 (Fourth edition, Sixteenth printing)

4. Winchell, A.N. and Winchell, H., The Microscopical Characters of Artificial Inorganic Solid Substances: Optical Properties of Artificial Minerals, Academic Press, New York, 1964

5. Pohl, F.H., A Field Guide to Rocks and Minerals, Houghton Mifflin Co., Boston, 1976

6. MacKenzie, W.S. and Adams, A.E., A Color Atlas of Rocks and Minerals in Thin Section, Wiley Halstead Press, New York, 1994

7. Mange, M.A. and Maurer, H.F.W., Heavy Minerals in Colour, Chapman & Hall, London, 1992

8. McCrone, W.C., McCrone, L.B. and Delly, J.G., Polarized Light Microscopy, McCrone Research Institute, Chicago, 1987

DEVELOPING A BETTER UNDERSTANDING OF BORON EMISSIONS FROM INDUSTRIAL GLASS FURNACES

Andrew Zamurs
Tim Batson
David Lever
Simon Cook
Suresh Donthu
Rio Tinto Minerals, Greenwood Village, CO 80111

ABSTRACT

Boron is volatile in the glass making process. In Europe, the second version of the reference document on Best Available Techniques (BAT reference, or BREF) for the manufacture of glass was adopted by the European Commission in February 2012 and published in March 2012. This revised version included an increased focus on boron emissions from furnaces. Thus, an understanding of the factors that affect boron emissions, and how they might be influenced, has never been more commercially relevant. This paper, presents laboratory data on how factors such as glass composition, combustion atmosphere, gas flow rate and temperature affect boron volatility. Based on similar data, predictive models are included that could help the glass industry better address boron emissions at the industrial plant level.

INTRODUCTION

Borates are an essential raw material for glass making. They provide benefits to the glass melting and forming processes by way of lower viscosity, lower energy requirements, and lower glass formation onset temperature. Some of the potential benefits borates create in the finished glass are decreased dielectric constant, lower thermal expansion, increased infrared (IR) absorption and increased chemical resistance.

While borates provide many benefits to glass, there are also potential issues that are associated with borates. Boron, along with sodium, sulfur, fluorine, and potassium, are among several elements that may volatilize in the glass manufacturing process. The importance of developing a better understanding of boron volatility is of interest from both an economic standpoint and from a rapidly emerging regulatory focus. Minimizing losses has a direct financial benefit that improves profitability of the glass manufacturing process. Additionally, in February of 2012, the second version of the reference document on Best Available Techniques (BAT reference, or BREF) for the manufacture of glass was adopted by the European Commission. This revised version included an increased focus on boron emissions from glass and frit furnaces[1].

In order to better understand boron volatility and the variables that influence it, a laboratory method was developed and employed to examine how different variables affect volatility. Also, a industrial scale furnace CFD model was constructed to examine how glass composition affects the rate of volatility.

LABORATORY VOLATILE EMISSION ANALYSIS

A method to quantitatively measure volatile emissions in a laboratory setting was developed and used to examine the effect of gas temperature and flow rate on volatility. The method uses an in-house built apparatus consisting of a furnace and volatiles collection apparatus, whose details are described in more detail in Zamurs, A. et al[2]

The efficiency and reproducibility of the method are shown in Table I and Table II, respectively.

Table I: Collection Apparatus Efficacy Analysis

	Na mg	%	B mg	%
Initial Wt mg	1823	100	1726	100
Final Wt mg	1698	93	1651	96
Collected	87	5	43	2
Sum	1785	98	1694	98

Table II: Reproducibly of the volatile analysis

	mg element per g glass	
Repeat	B	Na
1	2.9	5.3
2	2.9	5.4
3	2.8	5.1

The method was used to study volatiles from a typical glass wool composition (5.8% B_2O_3, 0.9% K_2O, and 15.8% Na_2O) in the temperature range 1200°C to 1400°C and with gas flow rates between 5 & 12.5 liters/min. Due to the composition and method limitation only B_2O_3, K_2O, and Na_2O were examined. The results are shown in Figures 1, and 2.

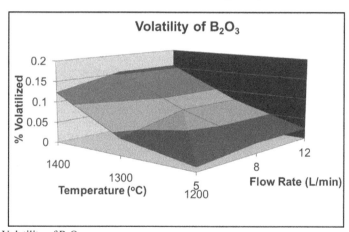

Figure 1. Volatility of B_2O_3

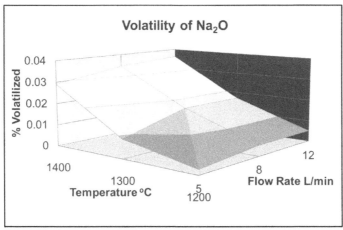

Figure 2. Volatility of Na$_2$O

As expected, temperature and volatility display Arrhenius behavior for all species because volatilization is thermally activated (Figure 3). The examining the slope of plot also indicates that the apparent activation energies for B$_2$O$_3$ and Na$_2$O are similar and that the energy for K$_2$O is slightly lower.

It was also determined that volatility reaches a maximum at a gas flow rate of approximately 8 liters/min (Figure 4). This is because at flow rates higher than 8 liters/min, the gas flow exceeds the rate at which volatile species can diffuse from the bulk melt to the surface, and react in order to become incorporated into the gas stream.

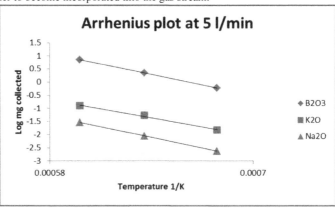

Figure 3. Effect of temperature on volatility

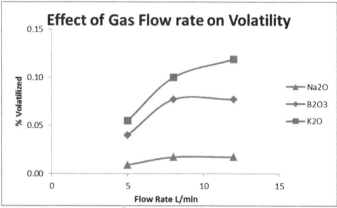

Figure 4. Effect of Gas Flow Rate at 1300°C

This effect may also be observed by comparing the apparent activation energies for a flow rate of 5 liters/min and 12.5 liters/min (Table III). At a flow rate of 5 liters/min, the rate determining step is the reactions at the surface of the glass and diffusion into the gas stream. At a 12.5 liters/min flow rate, the rate determining step becomes the rate that volatile species can diffuse through the molten glass to the surface.

Table III: Apparent Activation Energies

	Activation Energy kJ/mol		
Flow Rate	B_2O_3	K_2O	Na_2O
5 liters/min	110	95	111
12.5 liters/min	79	70	87

Next, the effects of the glass B_2O_3 content and MgO content were examined. The results are shown in Figures 5, 6, and 7.

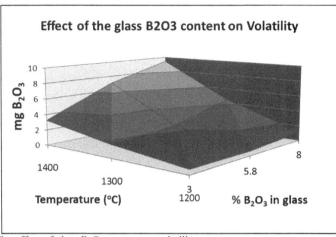

Figure 5. The effect of glass B_2O_3 content on volatility

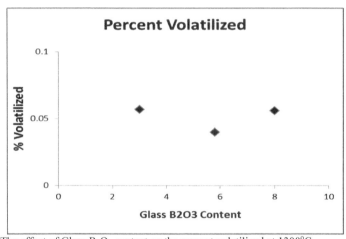

Figure 6. The effect of Glass B_2O_3 content on the percent volatilized at $1300^{\circ}C$

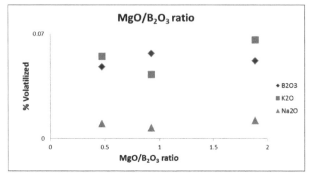

Figure 7. The effect of MgO/B$_2$O$_3$ on Volatility at 1300°C

When the effect of B$_2$O$_3$ content is examined, it is evident that total mass of captured B$_2$O$_3$ is greater at higher temperatures and higher B$_2$O$_3$ content (Figure 5). However, when total B2O3 content alone is examined, % B$_2$O$_3$ volatilized (fig 6) is more or less constant, suggesting this is primarily a temperature-driven phenomenon.

Reports in the literature suggest that MgO could have an effect on volatility. For example, Brow et al.[3] have suggested that MgO might have an effect on the polarizability of nonbridging oxygens which could in turn affect volatility. Based on the results shown in Figure 7 there appears to be no relationship between MgO/B$_2$O$_3$ ratio and volatility.

Further work should be done to examine the effects of similar variables, but on different glass compositions, e.g. the effect of MgO on alkali-free glasses. The effects of other oxides, such as CaO, BaO etc., should also be examined to see if there are any correlations with volatility.

FURNACE VOLATILTY MODEL

In order to further examine how the glass composition can affect volatility on an industrial glass furnace, Rio Tinto Minerals commissioned Celsian Glass and Solar to construct a furnace volatility model.

The model consists of a Computational Fluid Dynamics (CFD) furnace model combined with a semi-empirical volatility model based on experimental data. The resulting model can give valuable insight into the volatility rate of an oxy-fuel furnace and/or a recuperative furnace over a given range of glass compositions.

The graphs in figures 8 and 9 show how varying amounts of Na$_2$O and B$_2$O$_3$ in the glass composition affect the evaporation rate of NaBO$_2$. An interesting conclusion that appears when comparing the two furnace designs is an approximate five-fold increase in the evaporation rate for the recuperative furnace. The reason for this difference becomes apparent when the models of the melt surface are examined.

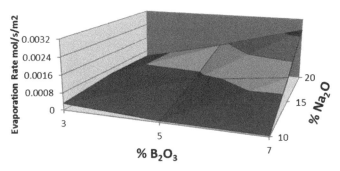

Figure 8. Volatility in an oxy fuel furnace

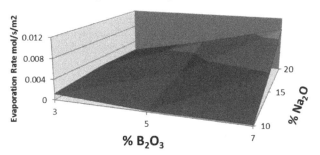

Figure 9. Volatility in an recuperative furnace

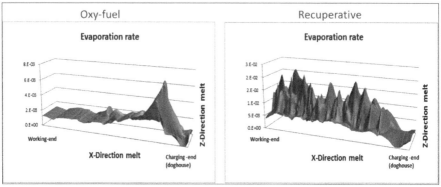

Figure 10. Models of the melt surfaces

The evaporation rate in the recuperative furnace is significantly higher at almost every point on the melt surface. The elevated evaporation rate in a recuperative furnace is due to much higher volumes of gas flow through the furnace, as a result of the air-fuel firing that this furnace design employs.

CONCLUSION

The possibility of emerging regulations regarding boron emission from glass furnaces enhances the value of understanding boron volatility. The experimental method and data presented here, along with the furnace volatility model, can improve our understanding of boron volatility and help the glass industry deal with any obstacles in the future.

REFERENCES

1. Joint Research Center – Institute for Prospective Technological Studies Sustainable Production and Consumption Unit. "Best Available Techniques (BAT) Reference Document for the Manufacture of Glass" Industrial Emission Directive 2010/75/EU
2. Zamurs et al. "The Influence of Borate Source on the Glass Forming Process" 73rd Conference on Glass Problems, Cincinnati, OH, 16-18 October 2012.
3. Brow et al. "Raman and 11B Nuclear Magnetic Resonance Spectroscopic Studies of Alkaline-Earth Lanthanoborate Glasses" *J. Am.. Cer. Soc.*, 79(9), 2410–2416 (1996).

NEW DEVELOPMENTS OF BATCH BRIQUETTING

Khaled Al Hamdan
Heiko Hessenkemper
Sven Wiltzsch
TU Bergakademie Freiberg, Institut für Keramik, Glas- und Baustofftechnik
Freiberg, Germany

ABSTRACT
New experiments in the laboratory and at a day furnace demonstrated that the grain size of the raw materials and the preparation of the batch can improve the melting behaviour of the batch, the homogeneity of the glass and reduces the evaporation / carryover of volatile components like Boron.

1. INTRODUCTION

The motivation for these experiments is to find a new strategy for the optimization of the glass melting process. These experiments have been founded in the course of the project „one step forming for the production of thin container glass". The optimization of the glass melting process comprised the improvement of the glass homogeneity, the specific melting rate, the avoidance of the dust, carry over, decomposition of the batch and prevention of the foam.

These goals can be achieved by granulation or compacting fine cullet together with fine raw materials. This can be realized by compaction the batch of fine raw materials and fine cullet and hydroxidic materials together.

2. LABORATORY TESTES

2.1 Compaction batch experiments

The advantages of the compaction are prevention of the batch separation, avoidance of the batch dust, increasing the heat conductivity of the batch and specific surface, which is exposed to the radiation of the furnace. For the compaction the desired properties of granules must be considered: optimal size 10 mm; narrow grain distribution; density $< 2,0$ g/cm^3, high mechanical strength; heatproof; homogenous; minimal humidity. For compaction the batch the following steps were taken:

a. Selection of the glass composition - The batch with fine cullet was prepared. The raw materials with the grain size distribution are shown in Table 1.

b. Characterisation of the raw materials - Grain size analysis has been conducted. The average grain size is shown in the Table 1.

Table 1. Middle grain size of the used raw materials

Raw material	D_{50} (mm)
Sand	0,2
Quartz powder	0,1
Dolomite	0,8
Soda	0,12
Lime stone	0,07
Cullet	0,2

c. The compaction agent and melting accelerators like NaOH, Ca(OH)$_2$, and K$_2$CO$_3$ have been used.

d. The granulation procedure was done with consideration of the sequenced mixing.

e. The humidity of the granules was determined.

Granulation tests were done in the granulation dish and granulation mixer.

The characterization of the granules is represented in the Tables 2 and 3 and Figure 1.

Table 2. Characteristics of the granules with cullet, created in the granulation Dish

Sample	bulk density g/cm3	compressive strength (MPa)	D mm	humidity % after granulation
Batch from natural raw materials	1,9	0,4	4-5	14,2
Batch from natural raw materials with CaO	1,8	1,1	6-8	12
Batch from fine raw materials	1,6	1,2	6-7	13,5
Batch from fine raw materials with 2% sodium silicate	1,9	1,8	9	12,7
Batch from fine raw materials with CaO	1,7	0,9	7-8	13,5
Batch from fine raw materials with(10 % Ca(OH)2)	1,6	1,2	5	11,70
Batch from fine raw materials with (10 % Ca(OH)2)	1,8	1,6	5	10
Batch from fine raw materials with (20 % NaOH)	1,7	2,5	6-8	10,8

Figure 1. Granules with cullet, created in the granulation dish

Table 3. Characteristics of the granules with fine cullet created in the intensive mixer

Batch	bulk density g / cm3	compressive strength (MPa)	D mm	humidity % after granulation
Batch from fine raw materials	1,8	1,1	< 7	10,6
Batch from fine raw materials with 50% CaO/ 50%Ca(OH)2	2,1	2,4	< 4	10,2
Batch from fine raw materials with 10 % Ca(OH)2	1,9	2,5	< 3	8,4
Batch from fine raw materials with 20 % NaOH	1,4	2,5	< 4	9
Batch from fine raw materials with 100 % NaOH	1,8	2,4	< 4	9,2

To increase the strength of the granules, the granules must be dried. To avoid the drying process the batch experiment with 3% humidity was pressed by a hydraulic compressor. The characteristic properties of the briquette are shown in the Table 4.

Table 4. Characteristic properties of the briquettes

Batch	compressive strength (MPa) (cold)	compressive strength (MPa) (warm 100°C)	density g/cm3
Batch from fine raw materials	30,5	38,9	1,9
Batch from fine raw materials with 50% CaO / 50% Ca(OH)2	45	44,9	2
Batch from fine raw materials with 10 % Ca(OH)2	45	44	1,95
Batch from fine raw materials with 20 % NaOH	31	42	1,95

The Table 4 shows that the addition of bonding agents like sodium silicate, NaOH, $Ca(OH)_2$ and heating of the NaOH enriched batch improves the compressive strength of the briquettes and granules.

Discussion of the granulation and pressing experiments

From the granulation (intensive mixer, granulation dish) and pressing experiments the following can be concluded:

- Granulation of the batch with cullet is possible in the granulation dish and in the intensive mixer with humidity 8 – 15 %.

- The addition of a bonding agent like NaOH, $Ca(OH)_2$ improves the compressive strength of the granules and reduces the humidity.

- The use of fine raw materials for granulation increases the compressive strength and helps to produce bigger granules.

- The grain size of the created granules in the intensive mixer is 1-3 mm and in the granulation dish, 5-8 mm.

- Briquetting the batch with 3% humidity by a hydraulic compressor or by a roller press and acceptable strength is possible.

2.2 Melting experiment

The laboratory tests were carried out with the following objectives:
- Solution of the foam problem
- Investigation of the influence of the decriptation on the briquettes
- Investigation of the influence of compaction on the meltability
- The influence of granulation and pressing on the homogeneity

2.2.1 Influence of the hydroxidic raw materials on the lifetime of the foam

The use of fine cullet and fine raw materials leads to the formation of foam. This Problem is solved by use of hydroxidic raw material. The hydroxidic materials have proven to be good granulation and acceleration agents. Further tests regarding their influence on the lifetime of the foam have been conducted by CCD - camera. In Figure 2 the influence of the hydroxidic raw materials on lifetime of the foam is shown.

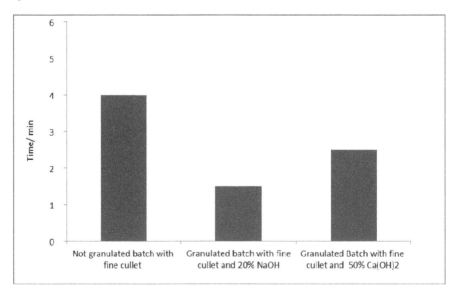

Figure 2. Influence of the hydroxidic Materials on lifetime of the Foam

The application of hydroxidic raw materials like NaOH is influenced positively the building of foam during the melting process of the batch contained fine cullet and fine raw materials, regardless of the initial contamination of carbon.

2.2.2 Investigation of the influence of the decriptation on the briquettes

The dolomite tends to spontaneous decriptation above 680°C by the melting of batch that leads to dramatically increase of the dust. The grain size range 100-500 μm is particularly susceptible to decriptation, see Figure 3 [Reference: lecture notes Prof. Conradt, RWTH Aachen].

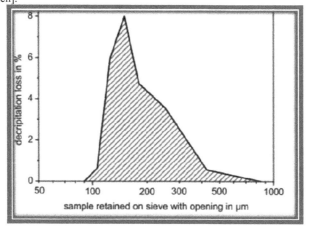

Figure 3. Decriptation dependent to grain size of dolomite

Compacted batch with coarse dolomite (< 500 μm) shows no spontaneous disruption in the hot stage microscope till 1200°C (our experiment).

2.2.3 Investigation of the influence of compaction on the melt behavior

The significant influence of the granulation and compaction of the batch with fine cullet on residue quartz dissolution time and refining are shown in the Figure 4.

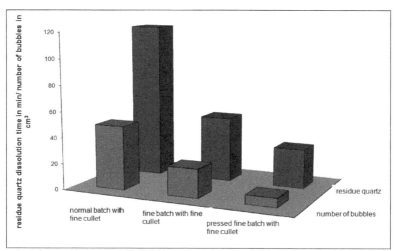

Figure 4. Relationship between the batch compaction and residue quartz dissolution time and refining

The determination of the residue quartz dissolution time in an electrical-heated tube furnace was done according to thread-pull procedures.

2.2.4 Melting experiments in special furnace
The melting experiments above are repeated in a special furnace (Figures 5 and 6). Two crucibles as like as boats were with the same speed into an electrically heated furnace in and out. The result of the special furnace confirms the tendencies in thread-pull procedures, see picture 3.

Granulated Batch

Reference Batch

Figure 5. Special furnace

Figure 6. Comparison between granulated batch glass and reference batch glass

2.2.5 Influence of granulation and pressing on the homogeneity
The prepared glass samples have been measured with Christiansen filter method regarding their homogeneity. From the homogeneity measurement of the glasses can be concluded that the homogeneity of the glass produced from fine raw materials and fine glass cullet is the better, see Figure 7.

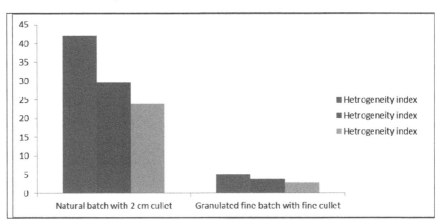

Figure 7. Influence of the cullet and grain size of the raw materials and the granulation on the homogeneity of the glass

3.   HALF-INDUSTERIAL TESTS

3.1 Compacting experiments by a roller press (Briquetting)
        The influence the compacting of flat glass batch and C-glass batch, grain size of sand and cullet on the melting behavior was tested. The compacting of the batches were done by a roller press. Figures 8 and 9 show a briquetted batch and a roller press, respectively. The flat glass batches contained a standard sand grain or quartz powder (with size < 0,1mm). The size of the flat glass cullet (20%) was less than 2 mm. The C-glass batches have quartz powder, too, but it contains no cullet and no standard sand grains.
        The half industrial compacting experiments of the glass batch showed that the compacting of the glass batch with a roller press is more economical, because the compacting needs not more than 3% water. In comparison to this the granulation on a granulating dish or intensive mixer needs at least 10% water, which means 7% higher energy demand to melt the glass batch. Furthermore, the roller press shows stable condition on the melting behavior of the glass melt and the properties of the briquettes. In contrary to this the process condition of the granulating dish or intensive mixer influence very much the properties of the pellets (granulate) and a lot the melting condition of the batch and could show melting problems.

Figure 8. Briquetted batch

Figure 9. Roller press

Discussion the Briquetting experiments
  • Briquetting of batch and cullet by a roller Press with higher strength is possible

  • Iron abrasion during briquetting by a roller Press caused no problem for mass glasses but for solar glass is still high, see Table 5, but can be reduced further by coatings of the roller.
  •
Table 5. Influence of Iron Abrasion during Briquetting by a Roller Press

Sample		not rolled	rolled
1 - 10	Fe_2O_3 %	0,024 ± 0,006	0,032 ± 0,008
1 - 10	Cr_2O_3 ppm	7,1 ± 0,005	11,3 ± 0,007

3.2 Influence of compacting on the melting behavior of the glass melt

The results of the experiments at the day furnace plant are presented in the Figures 10 and 11. The experiments at the day furnace plant show positive influence of the compacting on the melting behavior of the batch.

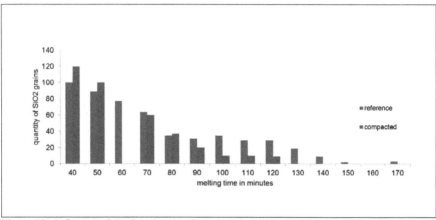

Figure 10.. Influence of the flat glass batch compaction on the residue quartz dissolution time

From Figure 10, it is obvious that compacted flat glass batch with quartz powder has a lower residue quartz dissolution time in comparison to a standard flat glass batch. In this case the melting efficiency could be improved around 10-20%. Figure 11 shows a dramatically reduction of the residue quartz dissolution time for a C-glass.

Taking into account that in both cases quartz powder were used then the effect could be explained by a decomposition of the standard batch. Due to this a stabilizing of the mixing condition of the batch with a compacting process is meaningful for the melting conditions.

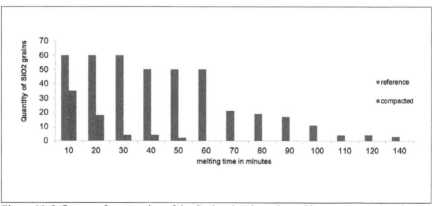

Figure 11..Influence of compacting of the C-glass batch on the residue quartz solution time

3.3 Influence of compacting on the evaporation of boron from the glass melt surface

Additionally, the batch dust (carryover) could be reduced by compacting of C- batch (Figure 12). Especially, the influence of the batch preparation (briquetting) on the carryover can be profitable for glass melts, which have a high loss of boron. Furthermore, the boron evaporation affected the lifetime of the furnace negatively. Costs due to lower furnace life, batch losses and less homogeneity of the glass could be saved, if the batch is compacted.

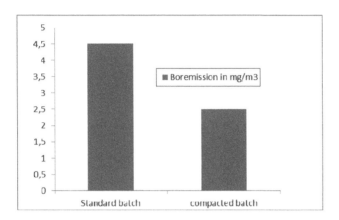

Figure 12. Influence of batch compacting on the boron emission

Additionally, a substantial decrease of the flue gas emission could be realized by the direct preheating of the briquettes. This seems possible, because the compacted batch should have fewer problems with caking in the batch-pre-heater.

SUMMARY

From past laboratory and the presented experiments in the day furnace can be concluded the advantages of compacting of the batch of fine raw materials and fine cullet as follows:

- Improvement of the glass homogeneity with possible effects to mechanical and optical properties
- Prevention of the batch separation
- Avoidance of dust and carry over in the furnace, regenerators and caking in the batch silo.
- Decrease of the foam life.
- Increase the melting rate of the furnace
- Reduction of the energy requirement
- Reduction of the evaporation of alkalis and volatile components such as boron
- Extension the life of the furnace
- Improving preheating technologies of the batch

OUTLOOK

The conversion of the batch compacting on an industrial scale is planned. The main aim is to improve the batch preheating and the melting behavior of the glass melt. It is expected that the energy efficiency of the glass production will be increased and the glass properties will be improved.

ACKNOWLEDGEMENTS

Thanks are given to the Federal Ministry for Education and Research (BMBF) and project executing organization Karlsruhe (PTKA) and our co-operation partners Heye International, Saint-Gobian Oberland, Maschinenfabrik Eirich, STG GmbH Cottbus, Otto Vision Technology and Waltec Maschinen GmbH. In addition, thanks are extended to our team, Mrs. Zschoge, Mrs. Glatz, Mrs. Voigt, Mr. Scheidhauer for their support performing laboratory experiments and chemical analyses respectively.

APPLICATION OF SELF-SUPPORTING PRECIOUS METAL STIRRERS IN THE
MELTING OF SODA-LIME GLASS

Alexander Fuchs, Applied Technology Glass Industry Solutions
Umicore AG & Co. KG, Hanau-Wolfgang, Germany

INTRODUCTION
For many years, self-supporting designs for glass stirrers made of grain stabilized PGM (Platinum Group Metals) alloys have proven their outstanding performance, primarily in specialty borosilicate glass applications. Until now, temperature homogenization in soda-lime glass applications was commonly achieved by using ceramic stirrers. Ceramic stirrers have historically been viewed as the most economical alternative for applications where stirring temperatures are usually lower and homogeneity requirements are moderate.

This paper aims at exploring the potential possibilities of lightweight self-supporting PGM-stirrers and their advantages when applied in soda-lime glass melting. Grain stabilized PGM-alloys, combined with advanced manufacturing methods enable the construction of efficient and long-lasting stirrers. In a market environment where glass requirements grow increasingly stringent, it can make technical (stirring efficiency) and economical (total cost) sense to replace or complement traditional stirring methods with grain stabilized PGM-stirring technology.

CERAMIC STIRRERS IN GLASS MELTS
Despite their comparatively low selling price, ceramic stirrers have several disadvantages. Like any other ceramic Zircon-Mullite, which is widely used as stirrer material, is sensitive to tensile- and shear stress. Its mechanical properties are also difficult to predict as the grainy material is anisotropic and cavities and cracks can occur. Inevitably this imposes severe limitations on the design of the stirrers in order to not exceed the low permissible stress of the ceramic. The stress constraint leads to very thick shaft diameters (>3" [80mm]) and bulky blades that must not protrude far from the shaft, shown in Figure 1. The homogenizing efficiency of a stirrer with small blades relative to a thick shaft is rather limited, especially when the stirrer speed must also be kept low to operate below the permissible torsion stress. Low stirring efficiency requires multiple stirrer arrangements multiplying the costs for stirrers and drives.

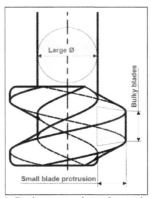

Figure 1: Design constraints of ceramic stirrers

Ceramic stirrers will be eroded by the glass melt, which in some applications can be very corrosive. Stirrer erosion leads to a reduction of stirrer efficiency in a relatively short time frame. In many applications the initial shape of the stirring blades is eroded almost completely down to the shaft within 12 weeks, as shown in Figure 2. With the stirring effect degrading both permanently and rapidly, glass homogeneity is hard to control. Consequently, the stirring behavior has to be constantly monitored and the stirrer speed has to be adjusted to compensate the wear. The eroded stirrer ceramic contaminates the glass melt and this can change the glass properties in undesired ways. Finally, eroded or corroded stirrer shafts can break off, notably at the three-phase zone along the glass line, causing major disruptions in glass production. The stirrer's relatively low selling price, quick and easy stirrer change may outweigh these disadvantages, especially when glass quality requirements are moderate.

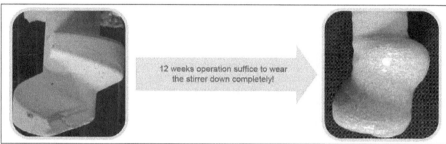

Figure 2: Ceramic stirrers degrade fast

ENHANCED GLASS QUALITY REQUIRES SUPERIOR STIRRING

Quality requirements for container glass are becoming more stringent in many applications. Improved temperature homogeneity of the glass gob is a key criterion for reproducible uniform thin walled containers with complex geometries. In many cases the number of gas blisters in the glass product must be reduced, and new furnace and channel designs now actively seek to reduce streaks and cord. Increased requirements for glass homogeneity and streak quality can exceed the mixing capabilities of traditional stirring technology.

Applying precious metals by coating or cladding the stirrer solves the most obvious problems of its ceramic counterpart:
- Resistance to corrosion and erosion, thus avoiding glass contamination
- Constant stirring behavior over time
- Higher temperature resistance
- Higher chemical resistance

Unfortunately, the low stirring efficiency that is ultimately dictated by the low tensile strength of the load-bearing ceramic is not improved by cladding or coating.

A self-supporting stirrer constructed with dispersion strengthened FKS® opens up a much wider range of possibilities and advantages. FKS® stands for Fine-Grain-Stabilized and is the Umicore brand name for their group of oxide dispersion strengthened precious metal alloys. A design using FKS® alloys can produce higher stirring efficiency by combining large and complex blades with a small diameter shaft. Smart design solutions and advanced manufacturing techniques make lightweight yet strong designs possible

As there is no material other than FKS® present in the hollow stirrer, it is insensitive to thermal shock and the danger of inducing electrochemical blisters into the glass is eliminated.

Recycling of a stirrer consisting of a single material is easier and thus metal recuperation rates are higher compared to coated or clad designs. Evacuation between the PGM-cladding and the supporting inner structure and the potential loss of vacuum is also not an issue in pure ODS stirrers. The service life of an FKS®-stirrer is significantly improved from 3 months in ceramics to 3 to 5 years depending on the application. Furthermore, evaporation of FKS® is much reduced compared to cast PGM alloy, thus reducing operation costs.

DISPERSION STRENGTHENED PRECIOUS METAL ALLOY FKS®.→ FROM CLADDING TO LOAD BEARING COMPONENT

Cast PtRh alloys subjected to load at very high temperatures will eventually creep and rupture. Material creep in a hot glass melt is a very different phenomenon to material rupture induced by exceeding the permissible tensile strength at room temperature. Material creep is the result of grain growth accelerated by high temperatures and slide occurrence along grain boundaries in the alloy's microstructure.

As an example, a 0.6mm thick PtRh10 cast alloy sheet which is exposed to 1,400°C for several weeks will eventually develop grains the size of the whole sheet thickness, shown in Figure 3. The large grains will slide easily and quickly along their boundaries even at stress levels lower than 5 MPa. This is why cast PtRh-alloys are applied only to coat or to clad a stirrer and not to carry load.

Figure 3: Micrograph of cast PGM-alloy showing grains as large as sheet thickness

Grain stabilized FKS® alloy uses oxide dispersion to inhibit grain growth and sustain the material strength throughout the lifetime of a stirrer. Umicore's FKS® materials have a creep resistance that is up to 1,000 times higher than the corresponding cast-alloy. The time to rupture of a FKS®Rigilit®PtRh10 sample (Figure 4) loaded with 8MPa at 1,400°C is 10,000 hours whereas PtRh10 cast alloy will rupture after only 10 hours. It is this outstanding material performance that enables the bearing of mechanical load, making the design of self-supporting stirrers possible.

Figure 4: Micrograph of FKS®Rigilit®PtRh10 (fine grain stabilized)

The resistance against grain growth is also illustrated by a laboratory test where an FKS®Rigilit®PtRh10 sample was charged with a tensile stress of 17MPa at a temperature of 1,400°C. The sample withstood the high load for 50 days (1,200h) before rupture. The micrograph of the rupture surface (Figure 5) clearly shows a brittle fracture and an otherwise fully intact fine grain structure. The dark patches are creep pores that are the starting points for material rupture. If that same sample is charged with half the tensile stress (8MPa instead of 17MPa) the time to rupture shifts from 1,200h to 10,000h (416 days) with the same grain structure and rupture appearance.

Figure 5: Micrograph of ruptured FKS®Rigilit®PtRh10
(Time to rupture: 1,200h @ 1400°C @ 17MPa tensile stress)

Creep-resistant FKS® helps to save precious metal, since wall thicknesses in the hollow stirrer can be reduced considerably. Because the whole stirrer is constructed with only one material, it has additional advantages:

- Precious metal recycling is not problematic and has a high metal recuperation rate.
- FKS® has a lower evaporation rate compared to cast PGM-alloy.

ADVANCED DESIGN AND MANUFACTURING TECHNIQUES TO FULLY EXPLOIT
THE FAVOURABLE FKS® PROPERTIES

Precious metal designs in FKS® will deliver excellent performance providing that several guidelines are adhered to. Using fusion welding disperses the material's fine grain structure locally. In other words, fusion welds have reduced creep properties and are potential weak spots. This problem can be successfully addressed by applying the following design rules:

- Avoid welding seams wherever possible
- Place necessary welds into areas of reduced stress
- Conduct forces and torques by interlocking connection to relieve stress in welds
- Welds are to be sealing as opposed to load bearing

Other design guidelines aim to save precious metal:

- Reduce material thickness in areas of reduced stress for economic use of precious metal
- Substitute precious metal with less valuable high temperature alloy wherever possible.

Adhering to these guidelines in practice requires special manufacturing techniques, such as:

- Tube drawing of seamless shafts
- Drawn shafts with stepped wall thicknesses
- Seamless spun parts
- Shrink fitted Inconel-couplers

The following example describes the transition of a clad molybdenum technical glass stirrer to a self-supporting stirrer solution and explains how these design principles and manufacturing techniques are put into practice. The starting point is state of the art PtRh10-clad molybdenum stirrer, shown in Figure 6.

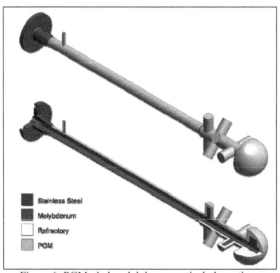

Figure 6: PGM-clad molybdenum optical glass stirrer

Cast PtRh10 is prone to creep, thus it relies on a Molybdenum core to transmit forces and torque. That core needs to be refractory coated, separating it from the PGM cladding to prevent inter-metallic diffusion. The volume between the core and the cladding must be evacuated to prevent molybdenum oxidation. The evacuation tube can be seen at the top of the stirrer shaft just below the connection flange.

The design using a combination of two metals results in additional requirements and a demanding and complicated layer structure of the stirrer. The cladding has a large number of welding seams in critical areas, and failure in these designs usually results from a crack in one of the welds, causing a loss of vacuum and ultimate failure of the stirrer after a service life between 1 and 3 years. The weight of the platinum cladding is also rather high, with a wall thickness of more than 1mm. Even though it is only a cladding it needs to have enough strength for the weld connections and to compensate for the differences in thermal expansion. The thermal expansion coefficient of molybdenum is only half of the coefficient of the platinum cladding, making the stirrer sensitive to thermal shock and requiring gradual heat-up before immersion in the glass melt.

Combining the unique properties of the FKS®-material with advanced manufacturing techniques enables the transformation to a self-supporting stirrer which eliminates all the complexity and disadvantages from the two-metal design. A good example of how to properly implement the ODS stirrer design principles is shown in Figure 7. The stirrer has a hollow seamless shaft, tubular blades that are also seamless, and a spun hemispherical bell at the bottom, so this design avoids welds wherever possible. Areas that have to be welded are sealing and not load bearing welds as shown in the section image at the bottom of Figure 7. Interlocking stiffeners carrying the load, which relieves the welds from stress, support each of the blades and the bell. Finally, the length of the precious metal shaft is reduced as much as possible, since the Inconel-super-alloy flanged coupler can withstand temperatures of over 600°C. Despite the fact that the stirrer is a self-supporting design, the metal weight in the FKS® stirrer is slightly lower than that of the cladding of the original molybdenum composite design.

Figure 7: Self-supporting optical glass stirrer

This stirrer's functionality is much improved, with a service life increased to five years, reducing the yearly costs for stirrer procurement and precious metal recycling. The design can transfer torques of up to 100 Nm at operating temperatures, has less evaporation losses than standard cast alloy, and is insensitive to thermal shock.

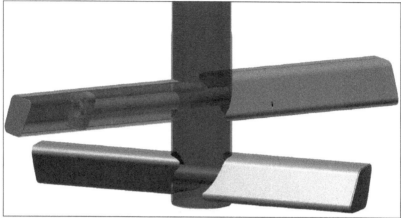

Figure 8: Example of a self-supporting channel stirrer for soda-lime glass

Figure 8 shows an example for a channel stirrer for a soda-lime glass application. The stirrer combines large stirrer blades with a sweep diameter of 350mm and a seamless hollow shaft with stepped wall thickness, 50mm in diameter and 700mm long. Interlocking torque transmission is executed with tubes that penetrate through the shaft and carry stiffeners at their ends to support the blades. With this support the welds connecting the blades to the shaft are only sealing and not load bearing and the resulting wall thickness can therefore be reduced considerably. The whole stirrer has a mass of 5,5kg FKS®Rigilit®PtRh10 and can permanently transmit a torque of 15Nm at 1300°C. The combination of the unique oxide dispersion strengthened FKS®-material with smart design solutions complemented with special manufacturing techniques result in very strong, yet lightweight stirrers.

TOTAL COST COMPARISON

Precious metal stirrers have higher manufacturing cost than their ceramic counterparts; however, when the total cost over the stirrer's lifetime is compared, the FKS®-stirrer with its increased functionality can make economic sense in many soda-lime applications. Several comparisons using a formalized total cost calculation template reveal a competitive case for the self-supporting precious metal stirrer. The biggest cost levers are longevity pushing down the yearly cost as well as functionality and increased operational safety.

The key cost factors to consider are:
- Longevity pushes the cost of ownership down significantly
- Economic use of precious metal
- Cost savings – less frequent stirrer changes
- Cost savings – less frequent ceramic stirrer procurement

- Cost savings – eliminate catastrophic stirrer failures
- Cost savings – higher glass yield, fewer glass defects
- Cost savings – high precious metal recuperation rates
- Cost savings – increased functionality may reduce number of stirrers and drives

CONCLUSION

Self-supporting precious metals stirrers are a proven and successful concept in the special glass industries. Self-supporting FKS®-stirrers are robust, provide years of reliable and constant service and allow complex geometries for high stirring efficiency. The unique material properties of FKS® make them lightweight, resistant to evaporation and fully recyclable with high metal recuperation. The technical advantages of this concept, with high potential for development, may be an answer to some of the challenges that the soda-lime glass industry is currently facing.

Glass Melting

APPLICATION OF AN ENERGY BALANCE MODEL FOR IMPROVING THE ENERGY EFFICIENCY OF GLASS MELTING FURNACES

Adriaan Lankhorst[1], Luuk Thielen[1], Johan van der Dennen[1], Miriam del Hoyo Arroyo[2]

[1] CelSian Glass & Solar B.V., Eindhoven, The Netherlands
[2] Universidad de Castilla-La Mancha, Ciudad Real, Spain

ABSTRACT

At CelSian Glass & Solar, a fast and flexible Energy Balance Model EBM for complete glass melting furnaces has been developed and validated. EBM is able to model energy balances for all furnace parts such as the melting tank, the combustion space, working ends, regenerators, recuperators and batch pre-heaters. EBM determines all heat flows and other interactions between these different furnace parts. It is equipped with an extensive database of batch, glass and refractory material properties. With this integrated furnace model, the overall energy balance as well as combustion, regenerator/recuperator and furnace efficiencies are determined. The total energy consumption in MJ/ton is determined, as well as structural losses and cooling losses. Crown and bottom temperature profiles are generated and flue gas and throat temperatures are reported. Thus, EBM enables the determination of the potential energy and CO_2 emission savings from which recommendations on modifications of process settings can be deduced and/or furnace designs can be optimized. The paper will show results of a validation study for a TV-glass furnace, for which detailed measurements are available. The same furnace has been used within the TC21 to validate several computational fluid dynamics (CFD) models against the measurements. Next, results of complete furnace simulations by means of EBM are presented, showing the effects on energy consumption of changing parameters such as cullet fraction, amount of gas, boosting heat, air excess, raw material composition, refractory insulation and cold air leakage. The impact of changes in these parameters will be discussed. The potential energy savings due to improvements as compared to the original design will demonstrate the applicability of EBM for optimizing the energy and CO_2 performance of glass melting furnaces.

INTRODUCTION

The prediction of the energy balance, heat flows and temperatures in complete glass furnaces (including heat exchangers) is very important for assessing the potential for reducing energy consumption, for finding avoidable energy losses and for comparing energy efficiencies of different furnace types. Such information can be obtained by performing detailed computational fluid dynamics (CFD) simulations, but these are generally time-consuming and usually only focus on specific parts of the glass melting process, such as the melting tank, the combustion space, the regenerator, etc. Elaborate industrial measurements in the furnace can deliver this information as well, but it is difficult to obtain a complete picture of the furnace energy balance from such measurements and it gives only information of a temporarily existing situation and not for furnaces to be built.

Therefore, at CelSian Glass & Solar, a fast and flexible Energy Balance Model EBM for complete glass melting furnaces has been developed and validated. EBM is able to model energy balances for all furnace parts such as the melting tank, the combustion space, working ends, regenerators, recuperators and batch pre-heaters. EBM determines all heat flows and other interactions between these different furnace parts. It is equipped with an extensive database of batch, glass and refractory material properties. With this integrated furnace model, the overall energy balance as well as combustion, regenerator/recuperator and furnace efficiencies are determined. The total energy consumption in MJ/ton is determined, as well as structural losses and cooling losses. Crown and bottom temperature profiles are generated and flue gas and throat temperatures are reported. Thus, EBM enables the determination of the potential energy and CO_2 emission savings from which recommendations on modifications of process settings can be deduced

VALIDATION OF THE ENERGY BALANCE MODEL

For a (former) regenerative sideport-fired furnace (see Figure 1), producing TV-panel glass at a pull rate of 236 ton/day, specific pull of 1.42 ton/m^2/day and cullet fraction 42.5%, detailed and accurate vertical glass temperature profile measurements have been performed by lowering a water-cooled thermocouple lance through the crown thermocouple blocks into the glass below. Besides this, crown temperatures, air preheat temperatures and flue gas temperatures have been measured, all at well-defined stable furnace conditions. The results of these measurements have been kindly made available by the glass manufacturer and have been used to define the Round Robin Test 5 (RRT5) for the ICG TC21 for validation of glass-dedicated CFD codes on the market [1]. The same measurements have been now used to validate EBM.

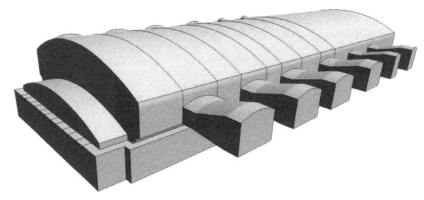

Figure 1: TV-panel furnace geometry

Figure 2 shows the positions (red indicators) on which the water-cooled lance was lowered into the glass. For the validation of EBM, the top glass surface temperature reading as well as the bottom temperature have been used, next to the crown, air preheat and flue gas temperatures.

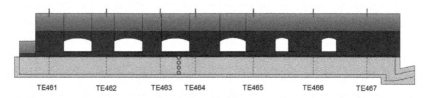

Figure 2: Location of depth temperature profiles

Typical operating conditions and other process-related data resulting from calculations by EBM, are shown below in Table 1.

Table 1: Summary of typical operating conditions of the furnace as calculated and reported by EBM

Pull rate (raw material)	264.3	[ton/day]
Pull rate (molten glass)	236.0	[ton/day]
Released batch gases	28.33	[ton/day]
Total area	166.6	$[m^2]$
Glass tank volume	176.9	$[m^3]$
Specific pull (based on raw material)	1.59	$[ton/m^2/day]$
Specific pull (based on molten glass)	1.42	$[ton/m^2/day]$
Reaction gas rate (volume)	769.3	$[nm^3/hr]$
Melting loss (relative to raw material)	10.72	[wt%]
Melting loss (relative to molten glass)	12.00	[wt%]
Cullet fraction (relative to raw materials)	37.88	[wt%]
Cullet fraction (relative to molten glass)	42.42	[wt%]
Average residence time	44.69	[hr]

The distribution of the amounts of gas and air ober the ports is shown in Figure 3. Only the first 4 ports are fired, with a fixed air-to-gas ratio. Port 1 receives the largest amount of gas in order to melt the batch, keep it from entering the bubbling zone and provide the required glass surface backflow which dives under the batch. On port 5 and 6, only cooling air is introduced, although preheated in the regenerators.

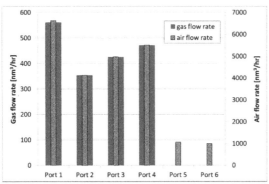

Figure 3: Fuel and air profile over the ports

In Figure 4 the measured and calculated crown temperatures are compared. For the majority of the crown, the calculated crown temperatures are within the uncertainty range of the temperature measurement, indicated by the error bars. In the refining area, where cooling air is applied on the ports, the crown temperature seems to be slightly under-predicted by EBM. However, it is reasonable to assume that the thermocouples in this cooling area show higher values due to the large amount of radiation they receive from the hot zone in the melting end.

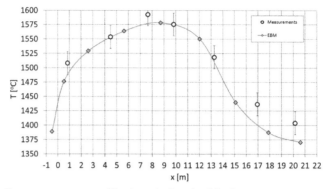

Figure 4: Crown temperature profile along the length of the furnace

From the vertical temperature profiles on the different locations (see Figure 2) the lowermost values have been used as the actual in-glass bottom temperatures. These are compared to the bottom temperatures as calculated by EBM along the length of the furnace in Figure 5. Due to the bubbling, which increases the recirculation and the amount of backflow in the melt, a relatively flat bottom temperature profile is observed. Over the entire length of the furnace, the agreement between calculated and measured temperatures is very good.

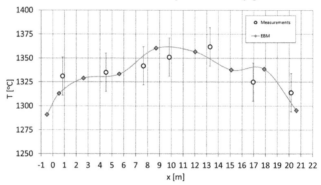

Figure 5: Glass bottom temperature profile along the length of the furnace

The uppermost temperature readings of the vertical temperature profiles are used to compare with the calculated glass surface temperatures (Figure 6). The glass hotspot temperature, around x = 8 m, is about 1 meter before the corresponding crown hotspot (see Figure 4). Over the entire length of the furnace, the agreement between calculated and measured temperatures is very good.

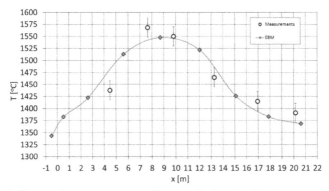

Figure 6: Glass surface temperature profile along the length of the furnace

In Figure 7 measured and calculated air preheat temperatures and flue gas temperatures are compared. Both air preheat temperatures and flue gas temperatures are measured in the top of the regenerators. Calculated and measured air and flue temperatures agree very well and are well within the measurement errors.

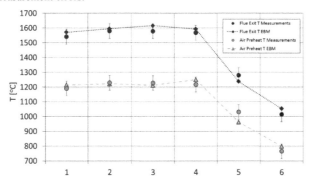

Figure 7: Air preheat temperatures exiting the regenerators and flue gas temperatures before entering the regenerators on the various ports

An overview of the most important EBM simulation results for the RRT5 TV glass furnace is shown below in Table 2. Characteristic temperatures are reported, typical environmental and energy consumption related data is reported such as total CO_2 emission (from fossil fuel and from batch), average O_2 in exhaust, specific fossil fuel consumption (in MJ/ton), efficiencies of all individual furnace parts as well as the overall furnace efficiency, structural heat losses, natural gas, air, batch gas and flue gas rates and finally costs related to fossil fuel (and electricity if used).

All temperature profiles show a very good agreement between the EBM model results and the measurements, indicating a good validation of the model.

Table 2: Tabulated simulation results of EBM for the regenerative TV-panel glass furnace.

Pull rate (molten glass)	236.0	[ton/day]
Fossil fuel consumption	7041.6	[MJ/ton]
Total heat flow to glass	3569.8	[MJ/ton]
Sensible heat gain glass	1484.5	[MJ/ton]
Throat temperature	1353.3	[°C]
Average flue gas temperature (reg's in)	1567.6	[°C]
Regenerator 1 average air preheat temperature	1241.0	[°C]
Regenerator 2 average air preheat temperature	1248.1	[°C]
Regenerator 3 average air preheat temperature	1237.9	[°C]
Regenerator 4 average air preheat temperature	1278.5	[°C]
Regenerator 5 average air preheat temperature	989.4	[°C]
Regenerator 6 average air preheat temperature	823.0	[°C]
Regenerator 1 average flue exit temperature	626.8	[°C]
Regenerator 2 average flue exit temperature	589.3	[°C]
Regenerator 3 average flue exit temperature	576.0	[°C]
Regenerator 4 average flue exit temperature	513.1	[°C]
Regenerator 5 average flue exit temperature	361.0	[°C]
Regenerator 6 average flue exit temperature	368.6	[°C]
CO_2 gas rate	483.2	[kg/ton]
Average O_2 concentration exhausts	4.76	[mole%]
Combustion efficiency	50.70	[%]
Furnace efficiency	21.08	[%]
Regenerator 1 efficiency	57.77	[%]
Regenerator 2 efficiency	59.78	[%]
Regenerator 3 efficiency	61.83	[%]
Regenerator 4 efficiency	64.89	[%]
Regenerator 5 efficiency	75.05	[%]
Regenerator 6 efficiency	72.58	[%]
Combustion space wall losses	1.61	[MW]
Glass tank wall losses	4.12	[MW]
Regenerator wall losses	0.55	[MW]
Total wall losses	6.28	[MW]
Fuel rate combustion	1810.3	[nm^3/hr]
Oxidant rate combustion	24518.0	[nm^3/hr]
Infiltration gas rate combustion	777.2	[nm^3/hr]
Flue gas rate combustion	27225.4	[nm^3/hr]
Fuel cost	21.72	[k€/day]
Fuel cost	92.05	[€/ton]

ENERGY BALANCE SIMULATIONS: ENDPORT-FIRED FURNACE

For an endport-fired furnace, producing clear soda-lime container glass at a pull rate of 220 ton/day and a specific pull of 2.07 ton/m^2/day, various energy consumption reduction measures have been investigated by use of EBM. The 10 year old furnace is on average operated at a cullet fraction of 42% and an additional 600 kW of electrical boosting is applied. Figure 8 shows an outline of the geometry of the furnace.

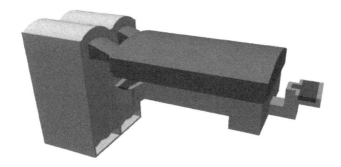

Figure 8: Endport-fired container glass furnace geometry

By means of parameter variations with EBM, the effect of amongst others air excess reduction, batch & cullet moisture content reduction and decrease of cold air entraining into the combustion space, have been investigated. The air excess is varied between 0% and 30%, the batch/cullet moisture between 0% and 5% and the cold air leakage into the combustion space between 0 and 2400 nm^3/hr. In all cases the throat temperature is maintained at 1350 °C. Typical compressed output results for the base case are shown in Table 3.

Table 3: Tabulated simulation results of EBM for the endport-fired furnace.

Pull rate (molten glass)	220.0	[ton/day]
Fossil fuel consumption	3748.9	[MJ/ton]
Boosting heat input	235.6	[MJ/ton]
Total energy consumption	3984.6	[MJ/ton]
Total heat flow to glass	2396.6	[MJ/ton]
Sensible heat gain glass	1654.5	[MJ/ton]
Throat temperature	1349.2	[°C]
Flue gas temperature (reg in)	1417.5	[°C]
Flue gas temperature (reg out)	413.7	[°C]
Air preheat temperature	1277.1	[°C]
CO$_2$ gas rate	312.5	[kg/ton]
Average O$_2$ concentration exhausts	1.92	[mole%]
Combustion efficiency	63.93	[%]
Furnace efficiency	41.52	[%]
Regenerator efficiency	68.39	[%]
Combustion space wall losses	0.97	[MW]
Glass tank wall losses	0.72	[MW]
Regenerator wall losses	0.47	[MW]
Total wall losses	2.16	[MW]
Fuel rate combustion	1087.3	[nm^3/hr]
Oxidant rate combustion	10069.8	[nm^3/hr]
Infiltration gas rate combustion	979.5	[nm^3/hr]
Flue gas rate combustion	12161.0	[nm^3/hr]
Fuel cost	13.05	[k€/day]
Electricity cost	1.35	[k€/day]
Total energy cost	14.40	[k€/day]

Figure 9 shows the effect of changes in air excess rate, batch & cullet moisture content and cold air leakage on the specific energy consumption. For this furnace, EBM shows that a specific fossil energy consumption reduction can be obtained of:
- 57 MJ/ton per 10% reduction of air excess (e.g. from air factor 1.20 to 1.10)
- 78 MJ/ton per % reduction of batch/cullet moisture (e.g. from 3% to 2%)
- 20 MJ/ton per 100 nm³/hr reduction of cold air leakage

Figure 9 also shows the effect of the same changes on the air preheat temperature after the regenerator. It is found that by decreasing the air excess, the air preheat temperature increases by 17 °C per 10% air excess reduction. A reduction of cold air leakage in the combustion space leads to an increase of air preheat temperature of 5 °C per 100 nm³/hr reduction. Although an increase in the batch/cullet moisture content leads to a significant increase in energy requirement, the air preheat temperature also increases, due to the increased flue gas rate (from 0 to 5% batch moisture, batch reaction gas rate doubles and flue gas rate increases by 16%) passing through the regenerator the energy exchange between flue and checkers becomes more efficient.

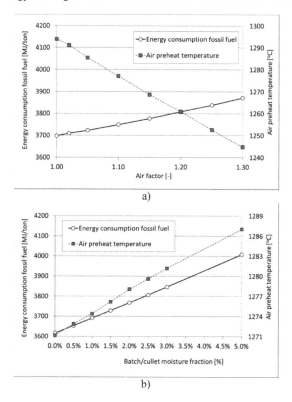

a)

b)

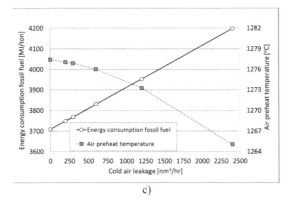

c)

Figure 9: Specific energy consumption and air preheat temperature as function of a) air factor, b) batch/cullet moisture fraction, and c) cold air leakage into the combustion space

Figure 10 shows the effect of the same variations on the regenerator, combustion and furnace efficiencies. When decreasing the air excess, the total amount of air and flue gases passing through the regenerator decreases, leading to a slight decrease in regenerator efficiency (defined as the ratio of the increase in sensible heat in the air to the total available sensible heat in the flue gas entering the regenerator). Both combustion space efficiency (defined as the percentage of the available combustion heat that is transferred to batch and glass) and furnace efficiency (defined as the increase in glass sensible heat divided by the total heat input) increase when decreasing the air factor.

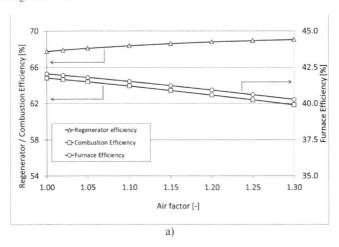

a)

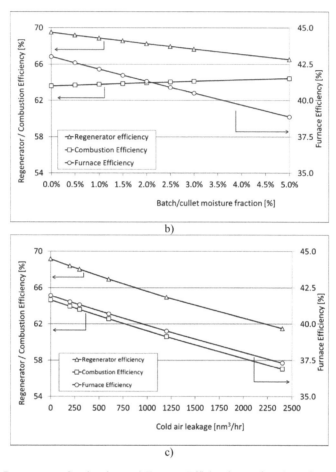

b)

c)

Figure 10: Regenerator, Combustion and Furnace Efficiencies as function of a) air factor, b) batch/cullet moisture fraction, and c) cold air leakage into the combustion space

Changing the batch/cullet moisture content does not significantly influence the combustion efficiency as the change in the combustion space radiative property is quite small. An increase in batch/cullet moisture content however leads to a significant decrease in furnace efficiency, due to the additional heat that is required to heat and evaporate the additional water in the batch. Although air preheat temperature increases for increasing batch moisture content, the regenerator efficiency (due to the nature of its definition) decreases as the amount of sensible heat available in the increased amount of flue gas increases. Reducing the amount of cold air leakage in the combustion space leads to an expected increase in all efficiencies.

ENERGY BALANCE SIMULATIONS: OXY-FUEL FURNACE

For an oxy-fired furnace, producing E-glass (see Figure 11) at a pull rate of 100 ton/day and a specific pull of 1.11 ton/m²/day, the use of different batch raw materials on furnace efficiency and energy consumption has been investigated by use of EBM. The 10 year old furnace is not utilizing cullet and an additional 750 kW of electrical boosting is applied. The effect of substituting either kaolin by anorthite and/or substituting limestone by burnt lime on energy consumption is investigated:

- Batch1: kaolin + limestone
- Batch2: anorthite + limestone
- Batch3: kaolin + burnt lime
- Batch4: anorthite + burnt lime

In all cases the throat temperature is maintained at 1430 °C.

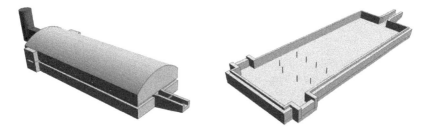

Figure 11: Oxy-fuel E-glass furnace geometry

Figure 12 shows the energy requirements, as calculated by EBM on the basis of its extensive raw materials database, for the 4 different batch compositions. The individual components of the energy requirement that can be distinguished are: heating of the batch, heating of the reaction gases, melting reaction energies and evaporation of moisture from the batches. The total energy requirement for batch-to-glass conversion in case of batch4 is less than 70% of the total energy requirement for batch1.

Figure 13 shows the melting loss for the 4 different batch compositions. The melting loss, defined as the total weight of the batch reaction gases relative to the weight of the raw materials or relative to the weight of the molten glass, decreases dramatically when applying batch4 to only about 16% of the melting loss for batch1. This has a significant effect on the combustion atmosphere, especially under oxy-fired conditions.

Due to these large changes in batch-to-melt conversion energy requirement and melting loss, important energy savings can be achieved by converting from batch1 to batch4. The specific fossil fuel consumption can be reduced from 7198 to 5555 MJ/ton (-22%). The furnace efficiency increases from 20.7% to 26.5% and the CO_2 emission rate decreases from 580 to 310 kg/ton (-47%).

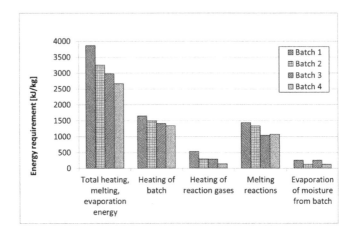

Figure 12: Energy requirements, as calculated by EBM, for the four different batch compositions: heating of the batch, heating of the reaction gases, melting reaction energies and evaporation of moisture from the batches

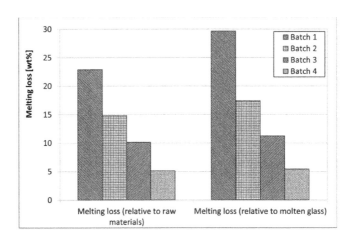

Figure 13: Melting loss (based on raw materials and on molten glass) for the four different batch compositions

Figure 14 presents crown and glass bottom temperature profiles along the length of the furnace for the 4 different batch compositions, as calculated by EBM. This graph shows that, besides important energy savings, additional operational advantages can be achieved by converting from batch1 to batch4:

- The bottom temperature beneath the batch increases by as much as 90 °C, leading to higher viscosities and thus to an increased back flow underneath the batch, resulting in a more efficient heat transfer from molten glass flow to batch
- The crown temperature in the batch area increases by about 30 °C, leading to a more flat crown profile and lower temperature gradients in the superstructure
- The crown hot spot temperature decreases by 20 °C, leading to an increased furnace life time and creating some room for increasing the pull rate

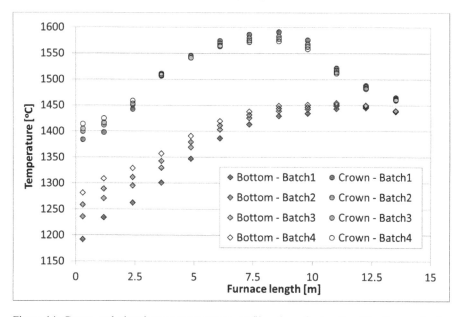

Figure 14: Crown and glass bottom temperature profiles along the length of the furnace for the four different batch compositions

Table 4 provides an overview of the potential energy savings in k€/day when converting from batch1 to batch2, batch3 or batch4. This table shows that, depending on local gas and oxygen prices, a reduction in energy cost can be achieved of as much as 2450 €/day when both kaolin and limestone are substituted by anorthite and burnt lime. However, depending on the price of the individual batch components, the net savings (taking into account the price of gas, electricity, oxygen and of the raw materials) can lead to a different conclusion: in the case shown in

Table 4 it proves to be most cost-efficient to only substitute limestone by burnt lime, leading to a net daily saving of 610 €/day.

Table 4: Energy savings and raw materials costs for the four different batches: evaluating optimal energy and raw material combination.

		batch 1	batch 2	savings batch 2 batch 1	batch 3	savings batch 3 batch 1	batch 4	savings batch 4 batch 1
Fuel rate combustion	[nm3/hr]	949	838	111	791	158	732	217
Oxidant rate combustion	[nm3/hr]	1805	1594	212	1506	300	1393	412
Fuel cost	[k€/day]	6.38	5.63	0.75	5.32	1.06	4.92	1.46
Oxidant cost	[k€/day]	4.33	3.82	0.51	3.61	0.72	3.34	0.99
Electricity cost	[k€/day]	1.2	1.2	0	1.2	0	1.2	0
Total energy cost	[k€/day]	11.91	10.65	1.26	10.13	1.78	9.46	2.45
sand	[k€/day]	0.95	1.20	-0.25	1.07	-0.12	1.31	-0.35
boric acid	[k€/day]	6.81	7.60	-0.79	7.88	-1.08	8.38	-1.58
kaolin	[k€/day]	3.70	0.00	3.70	4.28	-0.58	0.00	3.70
limestone	[k€/day]	2.71	1.95	0.76	0.00	2.71	0.00	2.71
CaO	[k€/day]	0.00	0.00	0.00	2.11	-2.11	1.45	-1.45
anorthite	[k€/day]	0.00	4.47	-4.47	0.00	0.00	4.96	-4.96
TOTAL	[k€/day]	14.17	15.22	-1.05	15.34	-1.17	16.11	-1.93
NET SAVINGS	[k€/day]			0.21		0.61		0.52

ON-LINE ENERGY BALANCE MONITORING

EBM can also be run in online mode, connected to a furnace. In this mode, it uses process data from the actual furnace (e.g. pull rate, cullet fraction, batch composition) as input parameters, to determine the current energy balance, temperatures and flow rates of gases. Output from the calculation can be visualized in a dedicated GUI. An example of a tab-page in such a GUI is shown in Figure 15, providing an overview of the actual energy flows in MJ/ton, flow rates in Nm^3/hr, temperatures and efficiencies of the furnace. The GUI is configured in close cooperation with the customer, resulting in dedicated visualizations of the important parameters to meet the requirements and wishes of the customer.

In parallel to the on-line calculation showing the actual status of the furnace an optimal case of the furnace for the current settings (e.g.no air leakage in the furnace, maximum crown temperature, pre-defined O_2 % in the flue gases) is also continuously calculated. By comparing the actual furnace operation with this most energy efficient situation for this specific furnace, which meets various constraints, the potential energy savings at any given time can be monitored, as well as the potential cost savings when changing the operating conditions of the actual furnace to be as close as possible to the optimal situation. Additionally, a "snap-shot" of the actual situation can be saved and be used off-line to study various energy savings strategies and evaluate and compare what-if scenarios.

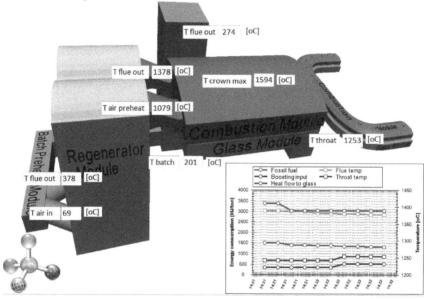

Figure 15: Visualization of output of EBM in on-line mode.

CONCLUSIONS

CelSian Glass & Solar B.V. has developed a fast and flexible Energy Balance Model (EBM). With this model the energy balance of an entire industrial glass melting furnace, including the air pre-heater and batch-preheater, can be calculated within a cpu-time of the order of minutes. Additional calculated information includes heat losses (structural, cooling), efficiencies and typical temperatures (throat, flue) and temperature profiles (e.g. crown). EBM thus provides a means for determining potential energy and CO_2 savings measures and EBM can be used to assess their impact before implementation. The model has been validated for the regenerative side-port fired furnace of the Round Robin Test 5 (RRT5) for the ICG TC21.

An example application of the EBM for an endport-fired furnace showed the impact of various changes in process parameters on the specific energy consumption and indicated the potential energy savings compared to the original design. A second example demonstrated the use of the EBM to evaluate energy savings of different batch recipes for an oxy-fired furnace.

EBM can also be run in online mode, providing an overview of the actual status of a furnace, quantify the possibilities for energy savings strategies and evaluate what-if scenarios. In conclusion, EBM is entirely accepted as an integral part of the design, operation, optimization and trouble-shooting of industrial glass melting furnaces.

ACKNOWLEDGEMENTS

We would like to thank Mathi Rongen and Martien Hendriks for their help in obtaining all required data to carry out the validation study. Special thanks to Frans van de Wiel of Philips Lighting BV for his determination to dive into the archives and dig up a lot of valuable data.

REFERENCES

1. A. Lankhorst, A. Habraken, M. Rongen, P. Simons and R. Beerkens, Modeling the Quality of Glass Melting Processes", 70th Conference on Glass Problems, Ceramic Engineering and Science Proceedings, Volume 31, Issue 1, 2010

OBSERVATION OF BATCH MELTING AND GLASS MELT FINING AND EVOLVED GAS
ANALYSIS

Penny Marson, Ruud Beerkens, Mathi Rongen
Presenting author: Ruud Beerkens
CelSian Glass and Solar, Zwaanstraat 1, 5651 CA Eindhoven, The Netherlands

ABSTRACT
    Several chemical reactions take place during the heating of a glass forming raw material
batch. These reactions are accompanied by a change in batch volume, the release of several
(batch) gases and occurrence of melting phases, gas bubbles and foam. Evolution of gas species,
such as CO, CO_2, O_2, and SO_2 can be monitored as a function of temperature. From this
information, the reaction mechanisms and temperatures can be derived during the fusion of the
batch into a molten glass.
    CelSian developed and uses experimental equipment to enable the observation of the
melting-in of a batch in transparent vitreous crucibles and simultaneous analysis of the evolved
gases from the batch during heating up, melting and fining. The furnace atmosphere during
melting and fining is controlled and the effect of different batch compositions on the temperature
of fining-onset and fining gas production can be measured. For instance, changes in the furnace
atmosphere can strongly influence the fining temperatures and fining efficiency.
    The equipment is applied to study the effect of:
- Batch pretreatment, such as grinding of the batch or pelletizing;
- Addition of cokes or different sulfate/coke ratios in the batch;
- Exchange one raw material for another type;
- Changing oxygen or water vapor level in the furnace atmosphere;
- Batch humidification;
- Addition of oxidizing agent;

on the melting-in behavior, foam formation and fining gas release. The experimental facility and
method will be described in this paper and a few relevant examples for industrial glass production
will be shown and discussed.
    This experimental method provides important information for glass technologists to
optimize the raw material chemistry and glass melting conditions to improve melting and fining
whilst limiting foam formation. This paper shows and discusses the application of this equipment
to examine the effect of batch preparation techniques (coarse, fine batch, pellets) on melting-in
and fining behavior.

INTRODUCTION
    Understanding the reactions which take place during raw material to glass conversion
gives insight into ways to adapt batch recipes or pre-treat the batch materials so that reactions are
able to take place in a favorable sequence such that less energy is required to make the glass.
In this paper combined Observation of Batch Melting and Evolved Gas Analysis is described as
used to understand the reactions, their sequence and kinetics for a simple soda-lime-silica glass
batch made from industrial grade raw materials. The Evolved Gas Analysis provides important
information on the occurrence of the reactions and the temperatures of fining gas releases from
the melt at different process conditions.

BATCH – GLASS MELT REACTIONS

During the initial melting of the batch several chemical reactions [1-3] take place. Many important reactions in the batch blanket or in the freshly formed silicate melt involve the formation of gases. The release of these gases depends on temperature, batch composition, furnace atmosphere composition and may be influenced by batch grain size or batch pretreatment.

When considering the most important reactions in a batch or melt with cokes and sulfate addition, cokes can directly or indirectly reduce part of the sulfates added as fining agents. This may result in release of SO_2 gas during melting-in of the batch or formation of sulfide components such as Na_2S [4-7].

An important example for many glass producers is the process of heating of soda-lime-dolomite-silica batches. These exhibit the following reactions:

1. Solid state reactions.

Solid state reactions between grains of different raw material species occur at the contact interfaces of the grains. The reactions finally result in the formation of silicates and release of reaction gases, for instance CO_2.

For example, reactive soda ash calcination by reactions with sand will cause CO_2 release and sodium silicate formation. In practice, this reaction mainly takes place above 780°C, when liquid phases are formed in the batch (eutectic melts of SiO_2 and sodium di-silicate: $Na_2O \cdot 2SiO_2$):

$$Na_2CO_3(s) + 2SiO_2(s) \quad \rightarrow Na_2Si_2O_5 + CO_2\uparrow \quad \text{(in practice at 780 - 900 °C)} \quad (1)$$

Solid state reactions are very slow. In practice reactive calcination of carbonates is often delayed by kinetic barriers and CO_2 evolution takes place at higher temperatures than expected from thermodynamic equilibrium calculations. Soda ash can further react with these sodium silicate melts, forming more CO_2.

2. Primary melt formation and melting of alkali rich carbonates.

The start of the melting process occurs in the temperature range: 700 - 900 °C. Typically significant formation of melts starts at about 800°C, in float glass or container glass SLS batches. Often this melting is associated with release of gases, due to reactions of carbonates and at reducing conditions or in CO atmospheres, reactions of sulfates forming respectively CO_2 and SO_2 gas species.

3. Dissociation or decomposition reactions.

Dissociation or decomposition reactions of Ca- and Mg- containing carbonates such as limestone and dolomite result in development of a large amount of CO_2-gas, by spontaneous calcination. A typical temperature range for these decomposition processes is 600 - 1000 °C, the lower temperatures for magnesium carbonate decomposition and the higher end of this range for limestone calcination.

For example decomposition of carbonates from dolomite

$$MgCO_3.CaCO_3(s) \quad \rightarrow MgO.CaCO_3 + CO_2\uparrow \quad \text{(in practice at 600 - 900 °C)} \quad (2)$$

$$CaCO_3 \quad \rightarrow \quad CaO + CO_2\uparrow \quad \text{(in practice at 750 - 900 °C)} \quad (3)$$

4. Dissolution of sand grains.

The sand reacts between about 750-1000°C with sodium silicates or soda to form liquid sodium silicates. Formation of the silicate-melt phases is associated with the release of CO_2-gas when limestone, soda or dolomite reacts with sand. Sand dissolves in - or reacts with - the primary phases. The temperature range for this process in practice starts at 750 °C and continues up to 1400°C. First soda ash and limestone may react and form a molten salt which subsequently reacts with sand to form silicates, and/or soda ash reacts directly with sand grains at the contact area between the grains or soda melt and sand. After most raw materials are converted into a molten state at temperatures of 950-1100°C, at further heating, sand grains dissolve in these melts. Sometimes this sand grain dissolution involves some gas formation (CO_2, SO_2).

5. Reaction of sulfur species

Firstly organic components in the batch, or present as a contaminant in cullet can form char upon the heating process of the batch and this char coke may react with the CO_2 originating from carbonate decomposition, according to the Boudouard reaction (4) in the temperature range from 700 to ±1000 °C:

$$C \text{ (cokes/char)} + CO_2 \text{ (from carbonates)} \rightarrow 2\,CO \qquad (4)$$

Reaction of sulfur species in the batch [5-8] and fining by decomposition or evaporation of fining agents dissolved in the molten glass [9]. Sulfates added to the batch can react with reducing species, such as cokes or CO gas, forming sulfides or SO_2 gas during batch heating (at about 700-1100°C), depending on the oxidation state of the batch according to (reactions 5-7) below.

$$2C + Na_2SO_4 \rightarrow 2CO_2 \text{ (gas)} + Na_2S \qquad (5)$$

$$4CO + Na_2SO_4 \rightarrow 4CO_2 \text{ (gas)} + Na_2S \qquad (6)$$

$$4C + Na_2SO_4 \rightarrow 4CO \text{ (gas)} + Na_2S \qquad (7)$$

Sulfides and sulfates may react with each other in the freshly molten glass to form SO_2 gas at 1140-1350°C.

$$Na_2S + 3Na_2SO_4 + kSiO_2 \rightarrow (Na_2O)_4 \cdot (SiO_2)_k + 4SO_2 \text{ (gas)} \qquad (8)$$

Spontaneous thermal decomposition of sulfates, often only observed for oxidized molten glasses, generally takes place at higher temperatures.

$$Na_2SO_{4(melt)} \rightarrow Na_2O_{(melt)} + SO_{2(gas)} + \tfrac{1}{2}\,O_{2(gas)} \qquad (9)$$

In some cases simultaneous CO_2 and SO_2 evolution can be observed in the temperature range 900 – 1100°C. These reactions will lead to sulfate loss without fining, because the temperature is still too low for fining and batch melting is not complete. Such reactions, at these temperature levels, can even lead to formation of larger bubbles in the viscous batch melts and eventually may cause batch foaming. Thus, this reaction will reduce the effectiveness of the fining process.

DESCRIPTION OF THE EXPERIMENTAL EQUIPMENT

In Figure 1, a scheme is shown of the experimental set-up, which was developed for the Evolved Gas Analysis (EGA) and simultaneous glass melting observation experiments. A constant carrier gas flow (in this study nitrogen gas, but other gases or gas mixtures can be applied) set by a mass flow controller is led into a tall vitreous silica crucible which contains the batch. The top part of the vitreous silica extends from the furnace and is sealed with a cooled lid. Two ceramic tubes pass through the lid and function as inlet and outlet tubes for the gases. Gases that are released from the batch mix with the carrier gas (in this case N_2) and are removed from the crucible space through the exit pipe and subsequently the gas mixture is led by heated pipes to the FTIR (Fourier Transform Infrared Spectrometer, MKS Model 2030).

Other gas mixtures are often used to simulate the furnace atmosphere in industrial melting processes, a mixture of water vapor, nitrogen, CO_2 and oxygen can be applied in such cases.

The volume fractions of several gases are measured by the FTIR, in particular the evolution of CO, CO_2 and SO_2. Gases that leave the FTIR are passed on to a cooler, to remove the water vapor. The resulting dry gas is passed on to zirconia oxygen analyzers which can measure concentrations of oxygen at different levels, on a volume % or volume ppm scale.

The experimental set up is also equipped with a high resolution camera. Frequent pictures are taken of the batch/glass melt in the vitreous silica crucible. These images are later used for, for example, detailed observation of the reactions taking place, onset of melting phase formation, measurement of foam amount, determination of the fining onset temperature. A film can be prepared from the recorded images. Comparing films from different batches shows clearly the effect of, for example, batch pre-treatment on the kinetics of batch to glass melt transformation or foaming tendency.

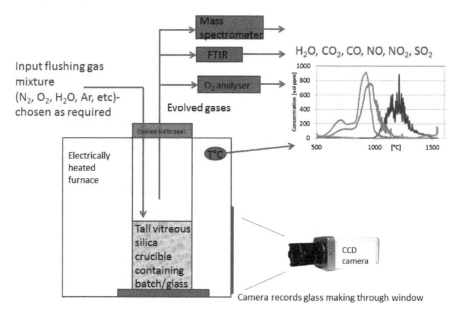

Figure 1: Schematic representation of the Evolved Gas Analysis and Melting Observation Set-up

Photographs of the assembly that constitutes the experimental set up are shown in Figures 2 and 3.

Figure 2: CelSian's Evolved Gas Analysis and High Temperature Melting Observation Apparatus

Figure 3: Evolved Gas Analysis Equipment containing FTIR and Oxygen analyzers

In this paper, we show experimental results to investigate the effect of batch pretreatment on gas release profiles from glass melt forming batches and on melting and fining onset.

In the experiments described in this paper, a batch mass of 100 grams was used. Glass batches composed of typical batch components used in the glass industry like, sand, soda, lime and dolomite with minor, but controlled additions of extra cokes and sodium sulfate were used. After melting, a glass with a composition typical for soda-lime-silica type flint container glass was obtained.

Here, the following temperature profile was applied:

- The transparent vitreous silica crucible with batch was placed in the furnace preheated at 500°C and the controlled amount of flushing gas was applied;
- 20 minutes was waited;
- The furnace was then heated from 500°C to 1550°C at a rate of 5°C/min;
- The furnace was kept at a constant temperature of 1550°C for 30 minutes

A slow heating rate of 5°C/min was selected to be able to relate the amount and type of gases that are released to a precise temperature interval at which the gas producing reaction takes place. In an industrial furnace, heating rates may be as high as 20 to more than 50°C/min.

The flow of nitrogen gas through the gas chamber of the crucible (the gas space above the melting batch) was controlled and the concentration of the different gases that were released was measured in the exiting gas flow. The total or cumulative quantities of gases that are released can be derived from the FTIR analysis and carrier gas volume flows. The amount of total actually released gas (CO_2, SO_2 and CO) can be compared with the expected quantity of the gas release, based on the batch recipe (carbonate and sulfate input) and sulfur retention of the glass.

EXAMPLE 1 THE EFFECT OF BATCH BRIQUETTES VERSUS NORMAL AND GROUND BATCHES ON THE MELTING-IN PROCESS

For this first example, 3 chemically identical batches were prepared. Two batches were ground in a ball mill to a fine powder (grain size < 50 μm). Briquettes (diameter 25 mm) were pressed (pressure of approximately 1500 Bar) from one of these fine batches.

Table I: Batch compositions for example 1: the effect of grinding and briquetting on melting-in.

Component	Mass %
Sand	63.5
Soda ash	18.96
Dolomite	17.14
Sodium sulphate	0.51
Cokes	0.03

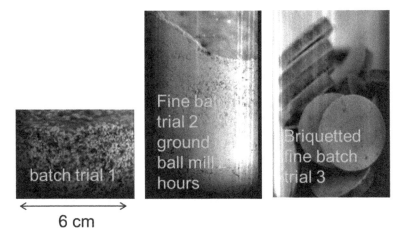

Figure 4: Photographs showing the batches used in example 1

The effect of the batch pre-treatment on melting-in reactions can be clearly seen looking at the evolution of carbon dioxide (from carbonate reactions) as shown in Figure 5 below.

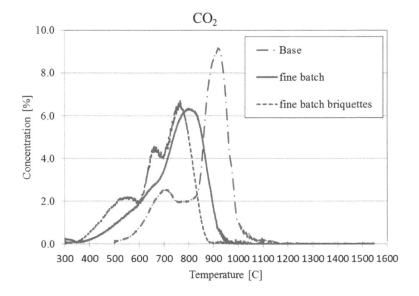

Figure 5: CO_2 evolution as a function of temperature

The graph shows that fine batch and pressing briquettes lowers the CO_2 peak temperature by 150 to 180°C: carbonates start to react at lower temperatures compared with the normal batch. Furthermore, the different CO_2 forming reactions (reactions 1-3) are better separated from each other on the temperature scale in the case of briquettes. Three peaks are visible with the briquettes prepared from fine batch. In the briquettes contact between the different raw materials has been optimal. This will enhance solid-state reactions in the batch. Grinding and briquetting is thought to enhance the carbonate-sand reactions, causing CO_2 release at a lower temperature. This would lead to energy savings in practice, because shifting endothermic reactions to lower temperatures will improve transfer of heat required for these reactions.

EXAMPLE 2: THE EFFECT OF COKE ADDITION ON FINING AND SO_2 RELEASE
        For this example 2 batches were prepared, one with (the standard batch) and the other without cokes.

Table II: Batch compositions for example 2: the effect of cokes on sulfur dioxide release and fining (removal of gas bubbles) behavior.

Component	"Base" mass %	"Base- no cokes" mass %
Sand	63.36	63.38
Soda ash	18.96	18.97
Dolomite	17.14	17.14
Sodium sulfate	0.51	0.51
Cokes	0.03	0.00

The effect of cokes addition can be seen clearly in the evolution curves for carbon monoxide and sulfur dioxide.

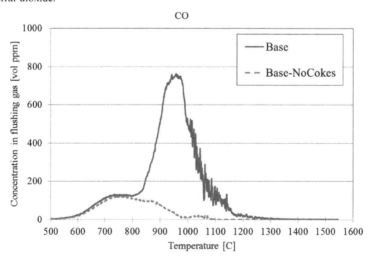

Figure 6: CO evolution as a function of temperature comparing batch with and without cokes added.

The CO release is caused by cokes-CO_2 reactions, during or just after carbonate decomposition reactions. The graphs show that there is a small amount of CO evolution from the batch even without cokes. This is probably due to the presence of some organic material in the batch. The amount of CO evolved for the cokes containing batch is attributed to the Boudouard reaction (reaction 4).

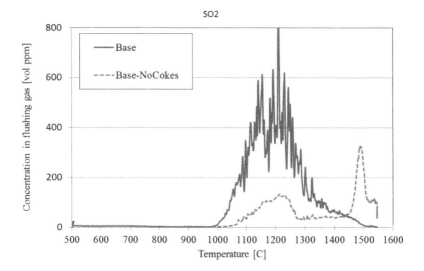

Figure 7: SO_2 evolution as a function of temperature for SLS batch with sulfate and with sulfate-cokes fining

The graphs show that the addition of a very small amount of cokes promotes an earlier and a larger release of sulfur dioxide. This is attributed to sulfate-sulfite reactions (5-8).
The higher SO_2 release peak (at higher temperatures 1450-1500 °C) for the no-cokes batch is attributed to the thermal decomposition reaction of sodium sulfate (reaction 9).
As less SO_2 is released in the non-cokes batch case, a correspondingly higher SO_3 retention can be expected in the glass. The SO_3 was measured for both glasses by ICP-ES as shown in the chart below.

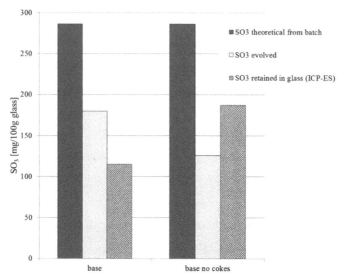

Figure 8: SO_3 retention in glass prepared from the batch with sulfate and cokes (Base) and the batch with sulfate without cokes (base no cokes).

EXAMPLE 3: THE EFFECT OF FURNACE ATMOSPHERE ON FINING ONSET TEMPERATURE

A typical float glass composition was melted from batch under different simulated furnace atmospheres, dry nitrogen versus simulated natural gas air firing and oxy-natural gas firing atmospheres. The different atmospheres contain a significantly different content (vapor pressure) of water. The graphs of sulfur dioxide release at high temperature (sulfate thermal decomposition) versus temperature are shown in the figure below.

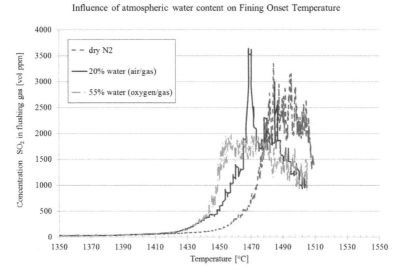

Figure 9: SO$_2$ evolution as a function of temperature

This graph shows that the presence of water vapor in the atmosphere reduces the fining onset temperature. The SO$_2$ evolution also increases more rapidly with higher atmospheric water content. The sulfur retention in the glasses prepared under water vapor rich atmospheres during melting are significantly lower than in sulfur analyzed in glasses prepared with lower water vapor contents.

CONCLUSIONS
Evolved Gas Analysis & High Temperature in-Situ Observation during batch-to-melt conversion provides a valuable method for many investigations. These can include investigating ways to melt-in faster by applying a certain batch grain size distribution, use of melting aids, using pellets, or other batch pre-treatment methods. Typical melting-in reactions are endothermic, if they are completed at lower temperature, great energy savings can be made.
New batch compositions can be studied to see the effects on foaming, melting-in & fining behaviour and to optimize a batch for fining.
The use of alternative raw materials can be explored, as can the effect of furnace atmosphere on fining onset (temperature) or foaming. The equipment is excellently suited to determine foam decay rates, or the fining onset temperature (for example to be used in CFD models to study fining behaviour in industrial tanks with certain glass melt flow patterns and temperature levels). Investigating of the effect of batch routing/selective batches or batch pre-treatment on melting & fining rates is another interesting application of these procedures.

REFERENCES

[1]   G. Nölle: *Technik der Glasherstellung*. VEB Deutscher Verlag f. Grund-
        stoffindustrie Leipzig German Democratic Republic 1979 ISBN 3 87144 494 4
[2]   J. Hlavac: *The Technology of Glass and Ceramics: An Introduction*. Glass Science and
        Technology **vol. 4** Elsevier Scientific Publishing Company,
        Amsterdam - Oxford - New York ISBN 0-444-99688-5 (vol.4) 1983
[3]   O. S. Verheijen: *Thermal and chemical behavior of glass forming batch*. PhD thesis
        Eindhoven University of Technology, 2003, 25. June, ISBN 90-386-2555-3
[4]   C. Flick; Nolle, G.: *Redox conditions during melting of the batch*. Glastech. Ber. Glass
        Sci. Technol. 68 (1995) , no. 3, pp. 81-83
[5]   P. R. Laimböck: *Foaming of Glass Melts*. PhD thesis Eindhoven University of
        Technology (1998) ISBN 90-386-0518-8
[6]   R. G. C. Beerkens.: *Sulphur chemistry and sulfate fining and foaming of glass*. Glass
        Technol.: Eur. J. Glass Sci. Technol. A, **48** (2007), no, 1, pp. 41-52
[7]   J. Klouzek; M. Arkosiová; L. Němec; P. Cincibusová: *The role of sulfur compounds in
        glass melting*. Glass Technol.: Eur. J. Glass Sci. Technol. A, (2007), **48**, no. 4, pp. 176-
        182
[8]   J. Collignon; M. Rongen; R. Beerkens: *Gas release during melting and fining of sulfur
        containing glasses*. Glass Technol.: Eur. J. Glass Sci. Technol. A, (2010), **51**, no. 3,
        pp.123-129
[9]   R. G. C. Beerkens: *Fining of Glass Melts: What we Know about Fining Processes Today*.
        Proceedings of the 69th Conference on Glass Problems, 4.-5. November 2008. Columbus
        OH, Am. Ceram. Soc. J. Wiley & Sons Inc. (2009) pp. 13-28

THERMOCHEMICAL RECUPERATION TO INCREASE GLASS FURNACE ENERGY EFFICIENCY

David Rue
Aleksandr Kozlov
Mark Khinkis
Harry Kurek
Gas Technology Institute
Des Plaines, IL, 60018

ABSTRACT

Glass melting is an energy intensive process that generates high temperature exhaust gas from either oxygen-natural gas or preheated air-gas flames. In air-fired furnaces much of the exhaust gas heat loss is returned to the furnaces using regenerators to preheat inlet air. Heat lost to the exhaust gases in oxygen-fired furnaces is rarely recovered. Thermochemical recuperation provides a means to use the heat in furnace exhaust gases to partially reform natural gas fuel. The hot syngas along with preheated air or oxygen is sent to the furnace burners. Modeling and demonstration testing have shown that TCR energy recovery is significantly higher than is possible with either thermal regenerators or heat exchangers. At the lower natural gas prices currently available in North America, simple payback times for glass melter TCR systems is calculated to be under four years.

INTRODUCTION

The quest to increase glass furnace energy efficiency has led engineers to develop improved furnace designs, to adopt better sensors and process control, to employ materials with superior thermal properties, to preheat batch and cullet, and to recover as much lost heat as possible by preheating inlet air.[1] Heat recovery is imperative for efficient air-fired furnace operation because of the large volume of high temperature exhaust gas produced. Engineered approaches for air-fired furnaces rely on air preheating using recuperators on unit melters and day tanks and regenerators[2] on large end port and side port furnaces. Commercial regenerators commonly heat inlet air to above 1100°C and lower exhaust gas heat loss to less than 30%. Returning more exhaust gas heat to an air-fired glass melter is not practical because of materials limitations and added regenerator cost. Exhaust gas heat loss from oxygen-fired glass melters is much lower than from air-fired melters and generally not recovered and returned to the melter because recovery costs do not justify equipment and operating costs with a short enough payback time.

Nearly a half-century ago, researchers in the USSR realized that heat recovery from hot exhaust gas streams could be enhanced by using the heat in the exhaust gas to partially reform the natural gas fuel to a furnace.[3-5] Developers learned that unlike the well-known catalytic reforming of natural gas to produce hydrogen that is carried out at high temperature and pressure, natural gas can be partially reformed at a much lower temperature and ambient pressure either with or without catalytic support.[6,7] A key discovery was that endothermic partial reforming at useful levels can be carried out if so-called thermo-chemical recuperation (TCR) residence times are long enough. The attraction of TCR is the potential to recover a larger fraction of the heat in the exhaust gas and thereby increase overall process efficiency. Researchers and developers in the USSR, Ukraine, UK, Japan, and USA have evaluated[3,4,8-11] and even installed[5,8,12] TCR units in multiple catalytic and non-catalytic configurations for many industrial processes and engines over the last several decades.

Natural gas is a mixture made up predominantly of methane along with a number of other alkane gases and small amounts of nitrogen and carbon dioxide. The reaction of methane with carbon dioxide and water at elevated temperature, known as reforming, produces a fuel gas with higher calorific value than methane. The primary methane reforming reactions are shown below.

$$CH_4 + H_2O\ (+Heat) \leftrightarrow CO + 3H_2 \qquad + 206\frac{kJ}{mole}$$
(1)

$$CO + H_2O\ (-Heat) \leftrightarrow CO_2 + H_2 \qquad - 41\frac{kJ}{mole}$$
(2)

$$CH_4 + 2H_2O\ (+Heat) \leftrightarrow CO_2 + H_2 \qquad + 165\frac{kJ}{mole}$$
(3)

$$CH_4 + CO_2\ (+Heat) \leftrightarrow 2CO_2 + 2H_2 \qquad + 247\frac{kJ}{mole}$$
(4)

The conversion of thermal energy to chemical energy is the mechanism used in the TCR process for recovering more heat in chemical form than is feasible to recover by heat exchangers such as recuperators and regenerators. TCR operates in a temperature regime under which only partial reforming occurs. The product gas from a TCR unit with exhaust gas and natural gas inlets is a mixture of a large fraction of methane combined with carbon monoxide, hydrogen, carbon dioxide, and water. The level of reaction completeness depends on reformer temperature, pressure, residence time, and unit physical configuration. As equations 1 through 4 show, TCR can be carried out with natural gas reacting either with carbon dioxide and/or steam. In practice this enables engineers to design TCR systems in which natural gas is mixed with exhaust gas containing carbon dioxide and steam or natural gas is mixed with steam alone. As will be shown, this flexibility provides TCR options that can be optimized for air-fired and oxygen-fired glass melters.

Figure 1 illustrates the application of the thermochemical recuperation principle. Heat recuperation and hydrocarbon fuel reforming is carried out in the unit shown as the Recuperative Reformer. In practice the Recuperative Reformer is often a multi-stage process arranged to maximize reforming level by managing temperature. The TCR process is also carefully designed to recover as much additional heat as possible through preheating of the reformed fuel gas and the air or oxygen to be sent to the furnace.

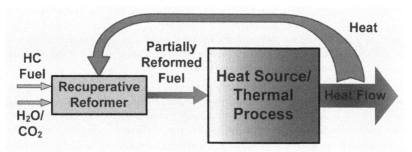

Figure 1. Thermochemical recuperation concept

THERMOCHEMICAL RECUPERATION FOR GLASS MELTING

In the early 1980s, engineers at the Institute of Gas Technology (IGT) and the American Schack Company explored the possibility of replacing air-fired glass melter regenerators with a thermochemical recuperation system. Engineering analysis and process modeling led to the development of the TCR system shown in Figure 2.

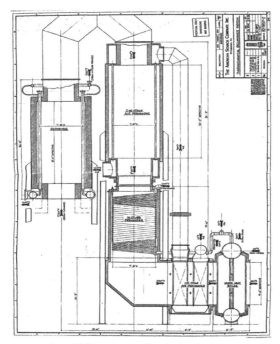

Figure 2. Steam-natural gas TCR concept for air-fired glass melters

In the final embodiment developed during this work, hot furnace exhaust gas travels through a radiative reformer, a secondary air preheater, a fuel gas mixture preheater, a first stage air preheater, and a waste heat steam generator. Natural gas is mixed with steam from the waste heat steam generator, and that mixture is heated in the fuel gas mixture heater. The heated steam-natural gas mixture is then heated further and reformed to a methane-rich syngas mixture in the radiative reformer and sent to the furnace burners. Furnace inlet air is preheated in the first stage and second stage air preheaters and then sent to the furnace burners.

The developed steam TCR concept for air-fired glass melters offered several advantages over regenerative heat recovery. First, exhaust gas temperature could be brought to a lower temperature and more heat returned to the furnace. This efficiency increase improved furnace economics but was not sufficient to warrant conversion to TCR. Other advantages included lower capital cost compared with regenerators, the elimination of the need to operate with furnace reversals, and the elimination of reversal valves and other hardware. Eliminating furnace reversals would allow for tighter furnace control and improved glass quality. A review of this thirty-year old design could prove promising considering the current costs of refractory and the need for lower emissions and tighter furnace control.

Oxy-gas glass melters operate with very hot exhaust gas that often contains corrosive components and particulates. Steam-gas TCR is again recommended so a clean fuel gas can be returned to the furnace burners. Unlike air-gas melter TCR, only the reformed fuel gas is returned to the furnace. Preheating the oxygen adds cost and complexity without providing significant efficiency increase. This simplification along with the smaller sized TCR equipment needed for an oxy-gas furnace lowers equipment and installation costs. Figure 3 shows a block flow layout of a basic TCR system for oxy-gas melters.

A concern with the TCR approach depicted in Figure 3 is the needed for water to generate steam needed in the reforming reaction. External water needs can be met internally by condensing water from the furnace exhaust gas. Figure 4 presents the layout of one possible configuration of an advanced TCR system for an oxy-gas melter requiring no inlet water.

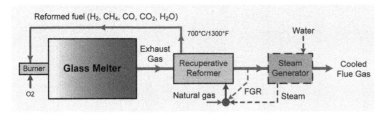

Figure 3. TCR for an oxy-gas glass melter

In the advanced TCR system, all particulates are collected and all water needed for steam reforming is provided through internal recycle. All process operations are carried out using proven steps. Scaling of TCR to the flow rates in oxy-gas melters is proven in other applications of the process steps. Endothermic thermochemical production of reformed fuel gas operates at lower temperatures than the temperatures in recuperators.

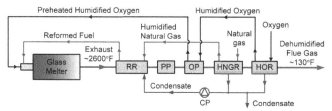

Figure 4. Advanced TCR for oxy-gas melters (CP – condensate pump, HOR – humidifying oxygen recuperator, HNGR – humidifying natural gas recuperator, OP – oxygen preheater, PP – particulate precipitator, RR – recuperative recuperator)

RESULTS AND DISCUSSION

Recent work has included modeling of the chemical and thermal balances around a representative oxy-gas glass melter and testing of TCR in a pilot scale process unit. The TCR modeling calculations are meant to be illustrative with the recognition that melter efficiencies and heat distributions are influenced by furnace size, temperature, refractory type, furnace age, and other factors. For the sample modeling case, an oxy-gas glass furnace was assumed to have an efficiency of 51.3%, with 29% of the input energy lost as exhaust gas at 1228°C (2242°F). The furnace gas phase material balance and overall energy balances are presented in Figure 5. Figure 6 presents the flow diagram and energy balance Sankey diagram for the same representative oxy-gas melter but with TCR included. A fuel savings of 16.4% is determined by modeling calculations. Furnace efficiency is calculated to increase from 51.3 to 61.4%.

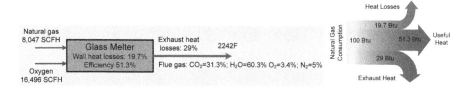

Figure 5. Energy balance for a representative oxy-gas glass melter

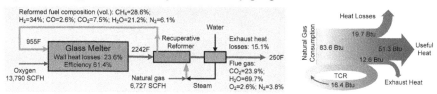

Figure 6. Energy balance for the representative oxy-gas glass melter with TCR

Installation of TCR will depend on payback time. Table 1 shows the expected TCR natural gas and oxygen cost savings based on several gas and oxygen supply costs for the same representative oxy-gas melter with a pull rate of 5 ton/h. Considering natural gas and oxygen cost savings alone, annual savings are found to be between $220,000 and $330,000. This is expected to equate to a simple payback time of 2 to 5 years. Payback times will be even shorter if natural gas prices return to historical levels of $8 per million Btu or greater. These cases were selected to correspond with current gas and oxygen costs for new furnace installations. Actual payback times are more difficult to estimate because prices can vary significantly and will not remain constant over the lifetime of a furnace. These cases serve at least as a baseline for assessment of TCR adoption.

Table 1. Project TCR Savings on a representative 5 ton/h oxy-glass glass melter

TCR System Savings Estimate -				
Natural gas - $/MCH	4	4	6	6
Oxygen - $/CCF	0.20	0.30	0.20	0.30
Natural Gas Savings per year, $	108,700	108,700	163,100	163,100
Oxygen savings per year, $	111,400	167,100	111,400	167,100
Total savings per year, $	220,100	275,800	274,500	330,200

The endothermic nature of the thermochemical recuperation process puts demands on the recuperative reforming process. Along with managing gas flows, providing proper surface area for heat exchange, and maintaining sufficient residence times, the design of a recuperative reformer must manage temperatures so the reforming reaction proceeds as desired. GTI engineers in partnership with experts at Thermal Transfer, Inc. developed a prototype recuperative reformer and then built a pilot-scale TCR test facility. The recuperative reformer approach is shown in Figure 7 and the pilot-scale TCR test facility is shown in Figure 8.

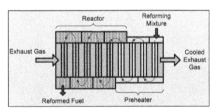

Figure 7. Pilot recuperative reformer

Figure 8. Pilot-scale TCR facility

The recuperative reformer is based on a tubular design. The unit built for the pilot-scale

TCR facility was designed for a maximum inlet exhaust gas temperature of 1040°C (1900°F), an exhaust gas flow rate up to 225 Nm³/h (8000 SCFH), and a reformed fuel gas rate up to 85 Nm³/h (3000 SCFH). The pilot-scale test facility was integrated with an existing laboratory furnace with a water-cooled load to simulate furnace loading. A Bloom burner modified for operation with hot reformed fuel gas was installed on the furnace and functioned flawlessly over the full range of trial conditions. The test system was operated over a wide range of TCR conditions and with a range of syngas to natural gas ratios. Reforming temperatures were as high as 760°C (1400°F). Combustion air temperature was varied from 15 to 650°C (60 to 1200°F). Combustion airflow was as high as 125 Nm³/h (4000 SCFH). Furnace energy input was as high as 150 kWh$_{th}$ (500,000 Btu/h). Flue gas recirculation was varied from 0 to 50%.

Equations (1), (3), and (4) describe the reactions of methane with steam and carbon dioxide. Trials were conducted by reforming methane mixed with an air-fired burner exhaust gas containing both carbon dioxide and steam. Hydrogen and carbon monoxide yields from the reversible reforming reactions are expected to rise with increasing temperature. Based on the chemical formulas, the ratio of hydrogen to carbon dioxide is expected to be between 2 and 3 based on how much methane reacts with steam and how much reacts with carbon dioxide. A series of conditions were established with a constant exhaust gas to natural gas ratio, similar residence times and flow rates, and multiple average reforming temperatures. There was no hydrogen or carbon monoxide in the exhaust gas when the recuperative reformer was bypassed and no reforming took place. Figure 9 shows the trend of increasing hydrogen and carbon monoxide with increasing average reforming temperature. At the highest temperature of 685°C, carbon monoxide and hydrogen were 10.5 and 25% of the reformed gas mixture. A H$_2$ to CO ratio of approximately 2.5 was observed at all reforming temperatures, confirming the same reforming reactions are taking place at all reforming temperatures.

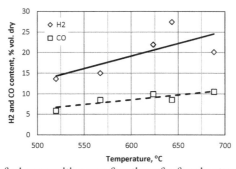

Figure 9. Reformed fuel composition as a function of reforming temperature

In other exhaust gas-natural gas TCR runs results were plotted to compare the furnace fuel savings achieved relative to modeling predictions and relative to a furnace operating with thermal recuperation returning 425°C (800°F) preheated air to the burners. The results of these TCR runs compared with the predicted model results and operating furnace are shown in Figure 10. Pilot-scale TCR results were slightly lower than modeling predictions but very close to the predicted values over the range of reforming temperatures tested. Modeling and the pilot-scale TCR testing showed reforming at these mild conditions produces methane conversions well below equilibrium values. Overall fuel savings from TCR were found to be 25 to 30% relative to a 425°C preheated air furnace with the highest fuel savings obtained at an average reforming temperature of 840°C (1550°F). The pilot-scale TCR runs were performed in duplicate, and conditions for each data point were held steady for several hours to assure stable reformer performance.

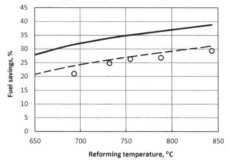

Figure 10. Fuel savings versus air recuperated furnace with 427°C air preheat: equilibrium estimation (solid line), prediction (dashed line), and test results (dots)

APPLICATIONS

The TCR process is compatible with a wide range of industrial processes and equipment. These fall into two broad classes. The first class is direct heat recovery through either partial steam or exhaust gas reforming. Examples of applications in this class include glass furnaces and installation on stoichiometric engines and steel industry equipment such as reheat furnaces. Large stationary engines are often used to produce electricity from low-calorific value landfill gas. TCR has been shown in GTI prediction to improve the operability and efficiency of these engines by using the excess heat in the engine exhaust gas to partially reform the landfill to a higher calorific value fuel gas that is returned at elevated temperature to the engine. The layout of an engine integrated with TCR is shown in Figure 11. In this catalytic TCR application the lower calorific value of the landfill gas containing 45% methane, 35% carbon dioxide, and 20% methane was increased by up to 30% and a hydrogen yield as high as 40% was obtained.

A second example of TCR direct heat recovery is for steel industry reheat furnaces. This application has been explored through modeling, design, and laboratory testing in work sponsored by the U.S. Department of Energy, the American Iron and Steel Institute, and the natural gas industry.

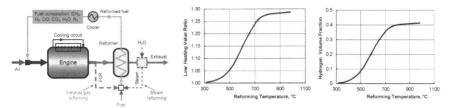

Figure 11. TCR Integrated With a Stationary Engine Along With Increases in Lower Calorific Value and Hydrogen Yield as a Function of Reforming Temperature

Figure 12 shows a conceptual application of TCR to a steel industry furnace and a design developed specifically for reheat furnaces. Maximum furnace efficiency increase is achieved for the reheat furnace with minimal TCR equipment. Exhaust gas leaving the furnace passes through a high temperature air recuperator, then through the recuperative reformed, and finally through a low temperature air preheater. Air is preheated in the low temperature air preheater, heated further in the air recuperator, and then sent to the burner. Natural gas is mixed with hot flue gas and then partially reformed in the recuperative reformer. The hot reformed fuel gas is then sent to the burners. Modeling estimates for the sample reheat furnace shown in Figure 12 an increase in thermal efficiency from 51 to 65%, a reduction in natural gas of 26%, a reduction in carbon footprint of 26%, and a reduction in NO_x of 30 to 40%.

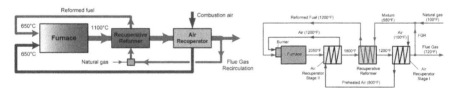

Figure 12. Concept of a Steel Furnace TCR System and a Practical TCR Implementation Approach for a Steel Reheat Furnace

The second class of TCR applications are those in which thermal energy is recovered in chemical form and the modification of the fuel gas by TCR is beneficial to the overall process. Examples of these processes include gasification processes, gas or coal to liquid processes, fuel cells, and solar-assisted natural gas upgrading.[13] Figure 13 shows one way in which solar energy-driven TCR can enhance natural gas fuel value. In this configuration a combustor can operate either with natural gas when no solar energy is available or with a hydrogen-rich fuel gas when the sun is shining. Ideal locations for this TCR application would be large industrial plants or power stations. Under optimum conditions the fuel gas calorific value can be increased by more than 20% with this increase dependent on the amount of solar energy that is available and that can be collected.

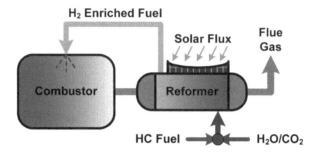

Figure 13. Solar assisted TCR

CONCLUSIONS

Thermochemical recuperation has been shown to be an approach to recovering additional energy from furnace exhaust gas than can be recovered by conventional thermal recuperation such as combustion air preheat or thermal regeneration. TCR relies on non-equilibrium partial reforming of methane and other hydrocarbons in natural gas with either steam or the carbon dioxide and steam in exhaust gas. Methane can be reformed with both carbon dioxide and steam or with the two gases together.

A number of applications exist in which TCR can provide significant process benefits. One group of applications includes recovery of furnace exhaust gas thermal energy in thermal and chemical form as a preheated syngas and the return of that heat to the furnace. Applications include steel industry furnaces and air-gas and oxy-gas glass melters, other high temperature furnaces, and stationary engines. In some situations where exhaust gases are non-corrosive, TCR involves reforming of a mixture of exhaust gas with natural gas. In other configurations, such as oxy-gas glass melters, where exhaust gases are either dusty or corrosive, steam is generated and a clean steam-natural gas mixture is reformed and returned to the melter as a hot syngas. TCR also offers the added potential when desired to produce all needed steam from exhaust gas condensation and to preheat inlet air.

The second class of TCR applications are those in which high temperature gases are used to recover energy along with the modification of the gas to produce other process benefits. Examples of these applications include solar heating to reform natural gas directly to produce a syngas with higher energy content, gasification and gas to liquids processes where TCR can modify the product syngas to achieve desired hydrogen to carbon monoxide ratios, and fuel cells.

Modeling and extensive pilot scale testing has shown TCR reactions to be well characterized. Despite the clear benefits of saving energy and reducing greenhouse gases, TCR has not yet been implemented industrially for several reasons. These hurdles of proof of concept are now being addressed with designs that are reliable and cost effective. Further demonstration testing to overcome development of robust industrial equipment is underway. These installations will provide the needed engineering information and cost information needed to proceed to the first commercial sales and large-scale adoption of TCR.

TCR offers a way to increase the efficiency of oxy-gas glass melters without modifying the melter or needing to adopt complex and costly batch and cullet preheating. Air-fired glass melters have also been shown to benefit from adoption of TCR as a replacement for regenerators. This application is not yet mature but deserves serious engineering review, modeling, and financial analysis.

ACKNOWLEDGEMENTS

The authors gratefully acknowledge support from the U.S. Department of Energy's Advanced Manufacturing Office, the California Energy Commission, gas industry Sustaining Membership (SMP) and Utilization Technology Development (UTD) programs, and Union Gas Ltd. Technology development partners and in-kind supporters have included the American Iron and Steel Institute (AISI), ArcelorMittal, U.S. Steel, Republic Steel, Bloom Engineering, Thermal Transfer Inc., Catacel, Cummins, and the Oak Ridge National Laboratory (ORNL).

REFERENCES

1. Limpt H., Beerkens R., and Habraken A. Overview of methods to recover energy from flue gases of glass furnaces – impact on glass furnace energy consumption. *Waste Heat Management in the Glass Industry*, WHM Workshop, Columbus, Ohio, October 21, 2010
2. Wachtman B., Nelson F. Waste heat recovery from regenerative glass furnaces using an air extraction process. *Proceedings Of the 51st Conference on Glass Problems: Ceramic Engineering and Science Proceedings*, Vol. 12, Issue 3/4, 2008
3. Novosel'sev V., Pereletov I., Khmel'nitskii R., et al. The thermochemical Recuperation of the heat in industrial high temperature processes, *Proceedings of the Conference on the Results of R&D work in Moscow Energy Institute in 1964-1965*. Moscow, USSR, 1965, pp.131-138.
4. Nosach V. Thermochemical Regeneration of heat. *Izvestiya Akademii Nauk USSR, Energetika i Transport*, Vol. 3, No. 5, 1987, pp.139-145 [in Russian].
5. Nosach V. *Fuel Energy*. Naukova Dumka Press, Kiev, 1989 [book in Russian].
6. Khinkis M., Kozlov A., and Kurek H. Non-catalytic recuperative reformer. *Patent application*. US 2012/0264986, Oct. 18, 2012.
7. Kweon C., Khinkis M., Nosach V., and Zabransky R. Advanced high efficiency, ultra-low emission, thermochemically recuperated reciprocated internal combustion engine. *Patent*, US 7,210,467.
8. Sikirica S., Kurek H., Kozlov A., and Khinkis M. Thermo-chemical recuperation. *Heat Treating Progress*, Vol. 7, No. 5, Aug. 2007, pp.28-31.
9. Maruoka N., Mizuochi T., Purwanto H., and Akiyama T. Feasibility Study for Recovering Waste Heat in the Steelmaking Industry Using a Chemical Recuperator, *ISIJ International*, Vol. 44, No. 2, 2004, pp.257-262.
10. Chakravarthy K, Daw S., Pihl J., and Conklin J. Study of the theoretical potential of thermochemical exhaust heat recuperation for internal combustion engines. *Energy Fuels*, No.24, 2010, pp.1529-1537.

11. Tsolakis A., Megaritis A., and Wyszynski M. Low Temperature Exhaust Gas Fuel Reforming of Diesel Fuel, *Fuel*, Vol. 83 (13), 2004, pp. 1837-1845.
12. Pratapas J., Kozlov A., Khinkis M., et al. Waste heat recuperation and conversion to fuel energy for higher efficiency and lower emissions from stationary natural gas engines. *Proceedings of International Gas Union Research Conference*, Paris, October 8-10, 2008.
13. Wegeng R.S., Palo D.R., Dagle R.A., Humble P.H., Lizarazo-Adarme J.A., Leith S.D., Pestak C.J., Qiu S., Boler B., Modrell J., McFadden G., "Development and Demonstration of a Prototype Solar Methane Reforming System for Thermochemical Energy Storage — Including Preliminary Shakedown Testing Results," 9th Annual International Energy Conversion Engineering Conference, July-August 2011.

DRY BATCH OPTIMIZER – GAIN ALL BENEFITS OF WATER-WETTING WHILE REDUCING THE DRAWBACKS

F. Philip Yu
Tom Hughes
Blaine Krause
NALCO Company
Naperville, IL 60563

ABSTRACT
The authors discuss the results and ROI of utilizing various surfactants to change the surface tension of water allowing the water to be used more effectively to "Wet" glass batch. Water-Wetting of glass is well known to have many benefits to glass production including reduced raw material segregation during transport, reduced dusting in furnace, improved furnace life, improved glass melting quality, etc. The drawback to wetting, however, is the cost of water and energy being used to evaporate the water in the furnace. In addition there are costs associated with poor mixing leading to batch feed equipment maintenance. The use of a small amount of specific surfactant has proven effective in reducing the amount of water to a minimum that is needed to effectively "Wet" glass batch while improving the homogeneous mixing of the batch. The result has been cleaner more efficient batch handling and feeding systems, more efficient consistent melt of batch, reduced water cost, reduced heat energy cost, reduced torque on batch mixers and subsequent electrical energy cost reduction, etc. This technology has already been implemented successfully at four sites in the US and is being trialed at two other locations globally.

BACKGROUND

Glass manufacturers use mixture of silica sand, soda ash, sulfate, limestone and dolomite as raw materials for batch processing. Additional cullet, at various ratios, is mixed with the dry batch before they enter the furnace. The batch recipe varies depending on the glass types, colors, and desired quality. Whether the dry batch is for float glass or glass packaging, the dry batch is melted into molten glass in a large furnace before coming out of the production line.

Glass companies typically add water, either via spray heads over the batch feeders or inside a large mixer, to maintain a target moisture level in the dry batch materials as they enter the furnace.[1,2] The thorough mixing of the dry batch will ensure a proper formation of batch piles as they enter the furnace to achieve melting efficiency and consistent glass quality. The wetting in the dry batch also contributes to reduction of dusting (carryover) as dry materials are added to furnace to prevent excessive particulate emission and fouling of exhaust systems /regeneration checkers. The accumulation of solids in furnace dome area, which is above molten glass level, can also be reduced with the proper wetting. The benefits also include minimizing damage to furnace refractory.

TROUBLESHOOTING GLASS DRY BATCH PROCESS
Onsite Troubleshooting:
A North American packaging glass plant used Nalco for cooling water treatment. The plant used a different vendor who provided a chemical wetting agent to improve the wetting consistency of the dry batch materials as they entered the furnace. The existing product contributed a buildup of organic material in the chemical feed tank and injection line, resulting in line pluggage and excessive maintenance cost. In additional to its excessive high cost, the product also contained sulfur compounds which led to SOx in the flue gas. Nalco, as the onsite

water treatment expert, was consulted to troubleshoot the problem and develop the most effective solution. After a thorough investigation in mechanical, operational and chemical aspects of this unique process, Nalco determined that the root cause was the bio-susceptibility of the existing wetting chemical.

A batch tank was used to make down the diluted wetting chemical before injection into the auger/screw feeder. The batch water tank, upon visual inspection, appeared to have heavy microbial fouling with slime formed around the edge of the tank. The differential microbiological analysis of the batch tank deposit (Table 1) showed diverse microbial population with majority of slime forming bacteria (*Pseudomonas*).

Table 1. Differential microbiological analysis of suspected biofoulants found in batch water tank of a packaging glass plant.

PHYSICAL APPEARANCE

Physical State	Solids	Quantity of Solids
Turbid liquid	Flocs	Heavy

Analyte	Result		Test Method
AEROBIC BACTERIA			CB22010
Total Viable Count @ 35°C	320000000 est.	CFU/mL	
Pigmented Bacteria	1 Type		
Mucoid Bacteria	Not Detected		
Total Coliforms	360000	CFU/mL	
E. coli	<100	CFU/mL	
Pseudomonas spp @ 35°C	380000 est.	CFU/mL	
ANAEROBIC BACTERIA			CB22016, CB22018
Sulfate Reducing bacteria	>100	CFU/mL	
FUNGI			CB22015
Mold	13000	CFU/mL	
Yeast	<10	CFU/mL	

Finding a Quick-Fix:

A follow-up laboratory screening was conducted using the plant's water to determine alternate products without any sulfur component. Two best performing products were then chosen to conduct trial runs in the plant. The replacement chemistry not only eliminated the fouling and pluggage issue at this plant, but also reduced the flue gas SO_X emission of the furnace. As an added benefit, the new wetting agent program (Product A) further improved the water contact with the dry batch. The shroud of the screw feeder showed improved cleanliness (Figure 1), which reduced equipment maintenance and replacement due to less abrasion of augers and dry batch feed equipment.

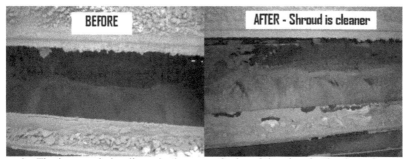

Figure 1. The improved cleanliness in the screw feeder of the glass bottle plant after treating with Product A.

Other benefits include reduction of carryover (dusting) in exhaust gases, leading to extended furnace life, less fouling of exhaust flues and environmental compliance (particle emission).

Follow-up R&D Work to Find the Optimal Solution:
The initial success in the glass process optimization had intrigued our interest to further study the interaction between dry batch materials and the wetting agent. The preliminary bench work was mainly focused on qualitative determination to show reduced sticking to sidewalls and improved fluffy nature of the dry batch. A laboratory simulation (Figure 2) of the screw feeder was set up to conduct quantitative measurement of the torque generated during the dry batch mixing process. A comprehensive screening was also initiated to evaluate other alternate chemicals to identify the product that generated the optimal performance. A Precision Stirrer with torque controller and measurement output was employed to drive an impeller assembly to simulate the screw feeder motion in the glass plant. The torque output was transmitted through an analog to digital signal converter to a data logger.

Figure 2. Laboratory setup (side and top views) to simulate dry batch screw feeder.

Dry batch mixture was maintained at 3-5% moisture level with various wetting chemicals to determine the product with the best performance on torque reduction. Product A was originally applied to resolve the pluggage problem at the packaging glass. However, a special feeding setup was required to handle the highly viscous product. Product B was later identified to provide improved performance over previous chemistry. Product B was applied in a sister plant of the original company, and showed improvement of consistency in product delivery with its lower viscosity.

The overall best performance was identified in Product C showing the highest torque reduction (Figure 3). With 3% moisture content in the dry batch mixture, Product C generated the lowest torque which would contribute to a smoother mixer operation and lower energy consumption. Product C was named the Dry Batch Optimizer (DBO), and the plan was to proceed with the third trial with the DBO program at the third plant.

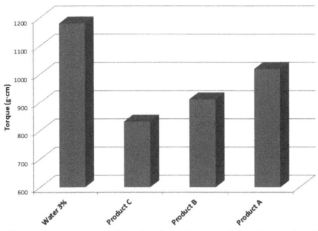

Figure 3. Torque measurement of dry batch mixing in a laboratory scale stirrer with double impellers to simulate screw feeder. All combinations were conducted under 3% moisture.

Redeploy the Dry Batch Optimizer:

From the troubleshooting the glass manufacturing process, we have concluded the following challenges often associated with typical batch wetting systems:

- In consistent distribution of moisture throughout the batch material, resulting in caking and plugging of wet-screws, chutes and chargers.
- Poor moisture distribution in batch affecting batch pile formation in the furnace (potentially affecting melting efficiency and quality)
- Adding extra moisture in order to obtain proper batch consistency (caused wasted energy)
- Inconsistent wetting, resulting in dusting, affecting particulate emissions and potentially shortening furnace refractory life.

A baseline data collection process was implemented to document both qualitative and quantitative data prior to applying Nalco DBO program. The Key Performance Indicators included:

- Moisture content (%)
- Cleanliness of wet screw and charger hopper
- Batch pile pattern quality

The third trial site was a new furnace just after being refurnished. Following baseline data collection, the DBO program was initiated, and the preliminary results in the early implementation stage was good. However, the plant considered the furnace was still being "stabilized" and would like to evaluate the impact of the DBO program to furnace operation by turning off the wetting agent. Within one of week of the Control-blank test stage, both the batch pattern and wet screw cleanliness became worse. In less than two weeks, the DBO program resumed feeding. (Figure 4)

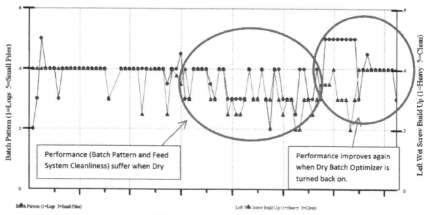

Figure 4. Qualitative observations of furnace operation at trial site #3 showing performance drop when Nalco DBO program was turned off. (Blue line/dot for Batch Pattern and green line/dot for left Wet Screw buildup; 1= bad and 5=good)

Over the next few months, the dosage was optimized and operators became more comfortable with the performance of DBO program. As good performance continued, operators were able to steadily reduce the moisture content in the dry batch operation and still maintained feed system performance and batch consistency in the furnace. The average moisture was reduced safely from 3.9% to 2.9% in six months as operators continued to fine tune the system. DBO made this possible by improving the way moisture distributed through the batch mixture, resulting in more even mixing (Figure 5).

Buildup of caked-on dry batch material in the feed chutes and hoppers often leads to flow restrictions and level control problems in a furnace. Prior to the implementation of Nalco DBO program, operators estimated that average 20 minutes per shift was spent on cleaning out feed system to prevent problems. During the seven months trial period, the dry batch chutes, augers and hoppers for the chargers of the furnace stayed considerably cleaner without significant intervention. Figure 6 showed clean dry batch feed system during inspection.

At trial site #3, the batch pile formation was reasonably good with just water as wetting agent (Figure 7a). However, the operators were challenged with adjustments to prevent "logging" and keeping good, distinct batch piles. The Nalco DBO program has maintained good batch pattern (Figure 7b), but with less adjustments necessary by the operators.

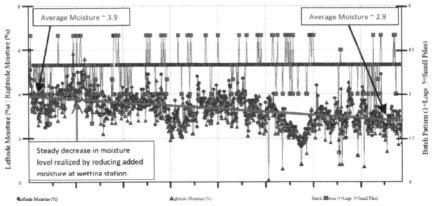

Figure 5. Dry batch moisture content was reduced by 25% from optimizing the Nalco DBO program in the furnace of trial site #3. (Blue: Left side Moisture; Green: Right side Moisture; Purple: Batch Pattern; 1=bad and 5=good)

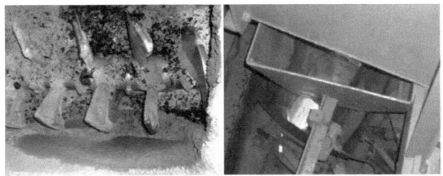

Figure 6. Clean screw feeder and charger chute observed during inspection at trail site #3.

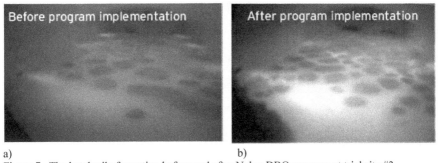

a)   b)

Figure 7. The batch pile formation before and after Nalco DBO program at trial site #3.

Preliminary Result at a China Glass Plant:

A float glass[3] manufacturer had agreed to run a short-term trial to test Nalco DBO program. There was not screw feeder in this float glass process. The dry batch materials were blended in a 3.6 ton mixer with a target moisture level and sent via a convey belt to the furnace. The cullet was added in the dry batch right before it reached the charge chute area. Due the buildup of dry batch residuals on the mixing blades, the mixer required routine maintenance every two weeks. In the winter season, the silica sand came with much higher moisture content and no water was regularly added in the dry batch mixer. Nalco DBO was applied to the dry batch mixer for a short period of time, and a significant improvement in mixer blade was noticeable from visual inspection (Figure 8).

 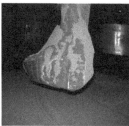 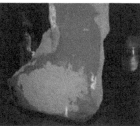

a) DBO   b) One week without treatment c) two weeks without treatment
Figure 8. The improved cleanliness in the dry batch mixer blade from a glass palnt trial site in China.

ECONOMIC RESULTS

The Nalco DBO program is designed to be added to the plant 's dry batch wetting water to improve the mixing between water and the dry batch component. The program has helped eliminate fouling and pluggage issue at trial site #1. Reducing SO_X emission in the furnace flue gas was an added benefit since the replacement product does not contain sulfur compound. We were able to address the customer's primary concern with its existing program while also reduced the chemical cost by $90,000 annually. In addition, there was a chemical cost avoidance of approximately $40,000/yr in biocide required if the plant continued to use the original program.

Among all three glass bottle plants, the DBO program has helped reduced the maintenance tasks such as cleaning screw feeders and unplugging chargers. The operators can concentrate on maintain overall furnace melting pattern and circulation to ensure efficiency and product quality.

The summary of Nalco DBO program benefits includes:

- Cleaner and more reliable feed system performance resulting in reduced intervention required from operators, realizing labor savings.
- Expected reduction of maintenance on batch feed system components due to less abrasion of screw feeders and dry mix feed equipment resulting in extended asset life.
- Ability to reduce batch moisture contents, without negative effect on production efficiency or product quality, resulting in reduced water and energy cost.
- Reduced excessive torque generated during incomplete mixing. The smooth operation of drive motor resulting in energy saving.
- Improve batch pile formation leading to improved performance consistency and glass quality.

REFERENCES

1. Davis, D. H. and Holy, C. J. To wet or not to wet – that is the question – part a, Ceramic Engineering and Science Proceedings (2011), 32.
2. Verheijen, O.S. Thermal and chemical behavior of glass forming batches, 2003 Dissertation of Technische Universiteit Einhoven, Netherland
3. Pecoraro, G. Method of making float glass. International Patent ZA7108362, 1973

Modeling, Sensing, and Control

IN-SITU CO AND O_2 LASER SENSOR FOR BURNER CONTROL IN GLASS FURNACES

A.J. Faber, M. van Kersbergen and H. van Limpt
CelSian Glass & Solar
Eindhoven, The Netherlands

ABSTRACT

A newly developed laser sensor system for measuring CO and O_2 in the burner ports of industrial glass furnaces is presented. Practical test results of the use of this sensor for monitoring the combustion process and for controlling the burners are given, for both an end-port fired regenerative furnace and an oxy-gas fired furnace. The use of the sensor for reducing NO_x emissions and improving the energy efficiency of glass furnaces is demonstrated.

INTRODUCTION

Nowadays, many glass companies use zirconia oxygen sensors for monitoring the combustion process and for adjusting the burner settings (air-fuel or oxygen-fuel ratio). However, the use of these solid state oxygen sensors in hot and corrosive flue gases has several drawbacks: the zirconia sensor provides a local oxygen concentration only, and often exhibits a relatively fast decline, resulting in false readings. Besides, usually no thermodynamic equilibrium exists between the different species in the hot flue gases of industrial glass furnaces, which will make the control of burners on just one, locally measured oxygen value, rather troublesome.

A newly developed laser sensor system, based on optical absorption measurements at specific wavelengths for CO- and O_2 -gas, respectively, does not have these disadvantages. This new sensor is non-invasive (no direct contact between sensor and flue gases) and provides average, representative values for the CO- and O_2 concentration over the cross section of flue gas channels.

In this paper, practical results of the use of this new sensor in both an oxy-gas fired furnace and a regenerative furnace are presented, demonstrating the use of the sensor for reducing NO_x and improving energy efficiency.

MEASUREMENTS IN END-PORT FURNACE

Figure 1 shows the configuration which was used for continuously measuring both the CO and the O_2 levels in the burner port of an end-port fired container glass furnace. In order to validate the measurement values, a water-cooled probe was used to extract a small volume of the flue gases from the top of the regenerator. During a period of more than 15 hours the concentrations of CO and O_2 in the extracted flue gases were measured using separate, calibrated gas analyzers. In figure 2 the CO-content (in vppm), measured in the burner port by the laser sensor is compared to the measured values in the extracted flue gases, showing a very good agreement.

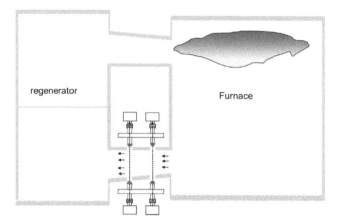

Figure 1: Measurement configuration of CO and O$_2$ laser sensors at the burner port of an end-port furnace

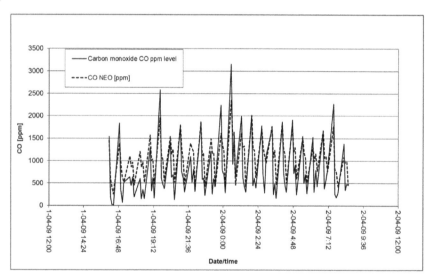

Figure 2: Comparison between CO values measured with calibrated gas analyzer ("Carbon monoxide") and signal of laser sensor ("CO NEO")

BURNER CONTROL ON THE BASIS OF CO IN END-PORT FURNACE

A representative measurement of CO in the hot flue gases provides a sensitive control parameter for adjusting the burner settings. In a test at the regenerative end-port furnace, the air-gas ratio of one of the two burners was controlled on the basis of the continuous CO-signal from the laser sensor, by a simple feed-back loop. Figure 3, showing the measured NO_x emissions versus the CO-set point, illustrates the results of this test. It is noticed that the oxygen concentration measured in the top of the regenerator for the setting with the lowest air excess to the burner (corresponding to around 4000 ppm CO in the burner port) was still 0.5 volume %.

Obviously, the NO_x emissions can be reduced to approximately 400 mg/sm_n^3 by decreasing the air-gas ratio of the burners, without going to very reducing combustion conditions. Another major advantage of minimizing the air excess to the burners is the possible saving on melting energy.

With help of an energy balance model, the estimated energy consumption (in GJ/ton of glass melt) of a typical 200-ton/day end-port furnace is calculated versus the air excess to the burners, see figure 4. It can be seen in this figure that by reducing the air excess from 9.4% (oxygen excess of 1.8%) to 2.4% (oxygen excess of 0.5 %), the estimated energy savings are more than 2%, corresponding to about 6 TJ/year.

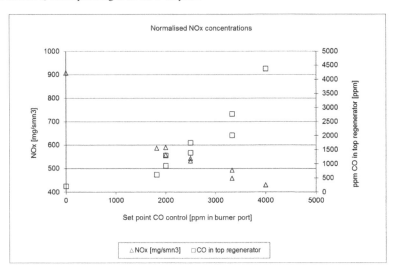

Figure 3: NO_x and CO concentrations in top of regenerator versus CO set point in control loop with laser sensor

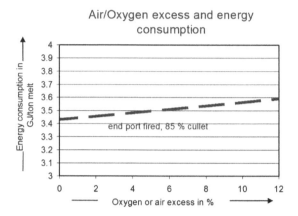

Figure 4: Estimated energy consumption versus air excess to the burners of a typical, 200 ton/day, and end-port fired glass furnace

MEASUREMENTS IN OXY-FIRED FURNACE

The measuring configuration used for testing a CO laser sensor in an oxy-gas fired container glass furnace, is presented in figure 5. In this latest configuration, laser and detector are both positioned at one side of the flue gas channel and a special mirror (retro reflector) is positioned at the other side. The advantages of this configuration include easier installation (no electronics in the hottest zone) and simpler optical alignment.

Again, extractive measurements of CO were compared to values measured with the laser sensor, showing a good mutual agreement. Moreover, an extensive measuring program was carried out for this furnace, during which the ratio oxygen-gas for the burners was varied systematically. The effects of varying the oxygen-gas ratio from around 1.85 – 1.86 to 1.83 on the concentrations of CO, SO_2 and NO_x in the hot flue gases, are presented in figure 6. Apparently, for this particular furnace, the NO_x emissions can be reduced to 0.16 kg/ton of glass, while still maintaining a small oxygen excess in the hot flue gases and without a noticeable increase in the SO_2 emissions.

On the basis of these measurements it is estimated that the oxygen-gas ratio can be decreased further, to about 1.81. This implies that the achievable savings on the oxygen consumption amount to about 1100 m^3 / day, corresponding to significant cost savings and savings on primary energy for the production the oxygen of around 2.5 TJ/year.

Meanwhile, this prototype CO laser sensor system has been in continuous operation for more than 1 year without interruption; the only maintenance required is a regular (2 weekly) manual cleaning of the mirror and the optical window.

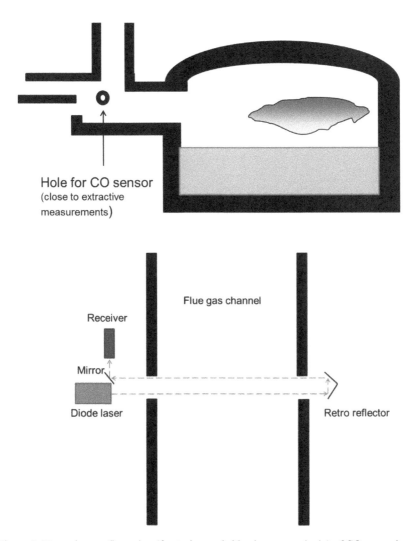

Figure 5: Measuring configuration (front view and side view respectively) of CO sensor in burner port of an oxy-gas fired glass furnace

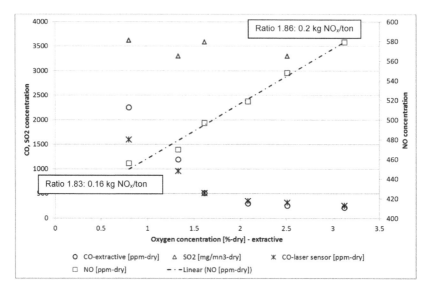

Figure 6: Effect of change in oxygen-gas net ratio (all burners) on CO, NO_x and SO_2 emissions in oxy-gas furnace

CONCLUDING REMARKS

It is concluded that a laser absorption sensor system for CO and O_2 provides a reliable and accurate method to monitor the combustion process and control the burners in both regenerative and oxy-fired glass furnaces. Industrial test results demonstrate the large potential of this new sensor type for decreasing NO_x emissions and improving the energy efficiency of glass furnaces.

It is envisaged that a final compact sensor design, comprising two lasers and two detectors (for CO and O_2) in one casing, will be commercially available by mid next year.

ACKNOWLEDGEMENT

The authors thank the company NEO (Norsk Elektro Optikk) for providing the sensor components for the tests

RADIATION IMPACT ON THE TWO-DIMENSIONAL MODELING OF GLASS SHEET SAGGING AND TEMPERING

Lochegnies Dominique[1,2], Béchet Fabien[1,2], Siedow Norbert[3], Moreau Philippe[1,2]

[1] PRES Université Lille Nord de France, F-59000 Lille, France,
[2] UVHC, TEMPO, F-59313 Valenciennes, France,
[3] Fraunhofer Institute for Industrial Mathematics, Fraunhofer-Platz 1, 67663 Kaiserslautern, Germany

ABSTRACT

Two-dimensional modeling of glass sheet sagging and tempering that includes solving the Radiative Transfer Equation (RTE) has been developed to take surface radiation, internal radiation and external radiative sources into consideration throughout the process. The RTE is solved using the P1-Approximation and is numerically implemented in ABAQUS® finite element software. The first original aspect of this study is that it solves the entire forming process, including forming and tempering as well as radiative effects. The second original aspect is that it solves the RTE on a non-fixed domain. The benefit of the method is demonstrated by comparing the deformed shape after forming and the residual stress after tempering by solving the RTE with P1-Approximation and by using approximate methods.

Keywords: Glass – Forming – Tempering – Radiation – Modeling

INTRODUCTION

During the forming and tempering processes, glass must be brought to a temperature over the glass-transition temperature T_g, which is about $823\ K$.[1,2] In forming processes, bringing glass to this temperature provides good deformability, thereby imparting glass with the final desired shape. During tempering, temperature level is essential to create a temperature gradient between the core and the surface of the glass as well as to enable stress relaxation to occur. At such temperatures, radiative effects become significant and cannot be neglected. Moreover, glass is a semitransparent material; consequently, radiative effects are present not only at the surface of the glass but also inside the glass.

In the glass forming and tempering literature, radiation is usually either neglected or approximated. A solution involves using the Stefan-Boltzmann law and considering only surface radiation[3]. This law is valid for opaque bodies. However, in the case of glass, surface radiation comes only from the opaque part of the glass. To get an accurate estimation, the solution presented in this paper is to integrate Planck function, which is different for each type of glass, to the opaque part of the spectrum. When internal radiation is taken into account, it is often done using the average of a modified conductivity comprising radiative effects. This can be equivalent to the conductivity obtained from experiments[4] or include the radiative conductivity computed with Rosseland approximation[5]. This method has the advantage of being very easy to implement. However, it is also known to be much too diffusive, especially for thin glass. Radiative sources (such as lasers) are generally considered a surface flux.[6,7] This is a good approximation if the laser wavelength is in the opaque part of the spectrum of the glass[7]. In contrast, if the laser is not in the opaque part of the spectrum, a part of the laser power goes directly into the glass or even through the glass.

A more accurate estimation of radiative effects is determined by solving the precise equation for determining radiative effects, known as the Radiative Transfer Equation (RTE).[8] This equation includes all the aforementioned radiative effects: surface radiation, internal radiation and radiative external sources. A detailed discussion of different numerical methods for solving the radiative transfer equation can be found in[9] and.[10] In the present study, this complex equation is solved to model sheet glass forming and tempering for the purposes of determining whether it is necessary to take radiation into account in a precise way by solving the RTE, or if approximate methods are sufficient.

TWO-DIMENSIONAL MODELING OF GLASS SAGGING AND TEMPERING

Forming and tempering operating conditions
The paper focuses on the sagging and tempering of a glass sheet (Figure 1) with the dimensions denoted $P = 1.5\ m$ for length, $L = 0.15\ m$ for width and $W = 0.006\ m$ for thickness respectively in the $\vec{x}, \vec{y}$ and $\vec{z}$ directions.

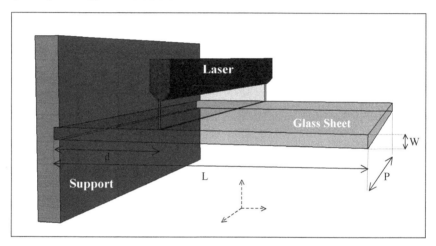

Figure 1. Three-dimensional description of glass sagging

The first step involved deforming the glass sheet. The glass was maintained at a temperature of $T_0 = 873\ K$ inside a furnace. In the modeling, it was assumed that the temperature inside the glass was homogeneous and equal to $873\ K$; at time $t = 0\ s$, a laser was applied for 25 s on line $y = d = 0.05\ m$ on the top surface of the sheet (Figure 1). The laser's power was $250\ kW.m^{-2}$ and the wavelength was between 2.75 and 4.5 μm. Due to the heat energy provided by the laser, the glass deformed by gravity sagging.

During the second step, the glass sheet was removed from the furnace and placed in the ambient air at a temperature equal to $293\ K$. Cool air with the same temperature was then blown on the sheet with a nozzle for $250\ s$. In the model, it was assumed that the glass sheet was moved out of the furnace instantaneously and the blowing was uniform over the glass sheet's entire surface.

Modeling
Considering the dimensions of the sheet $(P \gg L, W)$, the edge effects linked to surfaces $x = 0.\ m$ and $x = 1.5\ m$ can be neglected in the modeling and the same boundary conditions are present throughout the surface of the sheet when x varies from 0. to 1.5 m, since the problem

analysis can be limited to the two dimensions y and z. To take the modeling of dilatation in direction x into consideration, generalized plane strain conditions were used.

In both forming and tempering steps, a thermo-mechanical problem including large deformations and radiative effects had to be solved. Glass was modeled with viscoelastic behavior (Generalized Maxwell model). The strong temperature dependency of the behavior was taken into account with the "reduced time" concept and structural relaxation was taken into account with the "fictive temperature" concept. Thermal dilatation with structural relaxation effects was also considered[11]. With the high temperature-dependence of the mechanical behavior (Tables 1, 2 & 3 in the Appendix), the heat transfer in the glass equation must be solved by incorporating radiative effects into the forming and tempering steps to get accurate temperature fields for the entire process.

Glass is a semitransparent material. Therefore, at high temperatures, most of the heat transfer in the 2-D glass sheet (Figure 1) and heat exchanges at the boundary are accomplished through radiation (i.e., photon transport). This is described mathematically by the RTE, which is a high-dimensional transport equation[12]. Together with the heat transfer equation, one obtains a non-linear system of partial differential equations[10]. Moreover, in the current situation, where the glass sheet deforms by gravity sagging, the radiative heat transfer model must be solved on a deformable body, and consequently, a non-fixed domain. There is coupling between the mechanical equilibrium equations, the heat equation, and the radiative transfer equation.

Considering radiation during glass sagging and tempering

The radiation of the hot glass in the semitransparent wavelength region is described by the RTE[12]. If a band model with three wavelength bands for the absorption coefficient κ (Table 4 in the Appendix) is considered, then in addition to the heat transfer equation, a transport equation must be solved for each of the three wavelength bands[10].

Due to the coupling between the heat transfer equation and the RTE, the whole system used to compute the temperatures in the glass sheet during glass forming is high-dimensional, non-linear, and therefore, challenging to solve numerically. The P1-Approximation was used for the radiative part to obtain the radiative energy $G^k(\bar{x})$[10]. This approximation leads to three diffusion-like equations with Robin-type boundary conditions:

$$-\nabla \cdot \left(\frac{1}{3\kappa_k}\nabla G^k(\bar{x})\right) + \kappa_k G^k(\bar{x}) = 4\pi\kappa_k B^k\left(T(\bar{x},t)\right), \quad \text{in the glass sheet,} \tag{1}$$

$$\frac{1}{3\kappa_k}\frac{\partial G^k}{\partial n}(\bar{x}) = \frac{1}{2}\left(G_a^k(\bar{x}) - G^k(\bar{x})\right), \quad \text{at the glass sheet boundary,} \tag{2}$$

where $G_a^k(\bar{x})$ is a term depending on Planck's function $B^k(\cdot)$[9] for the surrounding temperature and on laser power $q_{laser}(t)$.

To solve the RTE using P1-Approximation, one can note that each of the three diffusion equations is very similar in form to the steady-state heat transfer equation. Not only does the equation have a similar form, but the boundary conditions do as well. Only term $\kappa_k G^k(\bar{x})$, present in the P1-Approximation (1) has no equivalent in the steady-state heat transfer equation. Consequently, the decision was made to solve the P1-Approximation using ABAQUS® finite element software because of its ability to solve the steady-state heat equation. The P1-approximation was solved using a DC2D8 thermal finite element of ABAQUS®(8-node quadrangle element with bilinear interpolation) modified with specific subroutines to correspond to equations (1-2).

MODELING RESULTS AND DISCUSSION
The results in terms of temperature changes during the process, of the deformed shape and of the residual stresses are presented for P1-Approximation denoted P1 and for an approximate method denoted Surface. In the Surface model, the laser is taken as a surface flux during the forming step; for the entire process, internal radiative effects are modeled with Rosseland approximation[5] and the surface radiation is computed with Stefan-Boltzmann law[3].

The thermo-mechanical problem was incrementally and iteratively solved using ABAQUS® finite element software. The 2-D mesh of the glass sheet (Figure 2) is composed of 8,987 nodes and 2,920 elements. At each time step, the Radiative Transfer Equation was solved first using modified DC2D8 elements to obtain, for each node, a radiative source term. Then, the thermo-mechanical problem was solved with CPEG8T elements (elements with generalized plane strain conditions, biquadratic interpolation for displacements and bilinear interpolation for temperatures) including this radiative source term in the thermal part.

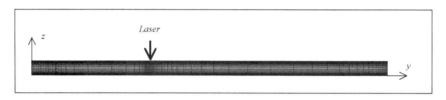

Figure 2. 2-D finite element mesh of the glass sheet in Figure 1

Results for the forming step
During the heating and forming step, the glass temperature evolved under the action of the laser. The natural convection is modeled with a film coefficient equal to $20 \ W.m^{-1}.K^{-1}$. Figure 3 gives, at different times, the temperature map of the deformed glass geometry obtained by P1 and Surface. The increase in temperature was mainly located under the laser; only one part of the sheet ($0.25m \leq y \leq 0.75 \ m$) is presented; the rest remained at $873K$ during the forming. Surface gave a more diffuse temperature field with the highest temperature values overall (Figure 3(b)). This can be explained by the fact that this model absorbs all the energy of the laser, whereas a part of it goes through the glass in reality and in P1 where the RTE is solved. After $25 \ s$ of heating and forming, the width of the heated zone is much greater for the Surface model (about $0.2 \ m$) than for the P1 model (about $0.1 \ m$). The explanation is related to the fact that Rosseland approximation is well-known as an overly diffusive method [12].

The deformed shapes of the glass sheet at the end of the forming step for P1 and Surface are presented in Figure 4. They were a direct consequence of the temperature field history in Figure 3. Considering the highest temperatures for Surface and the largest heat diffusion in comparison with P1, as just mentioned, the geometry change was clearly much greater for Surface. Deflection at the right extremity was respectively $17.2 \ mm$ for Surface and $8.9 \ mm$ for P1.

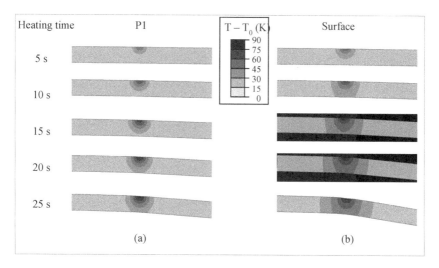

Figure 3. Temperature increase $DT = T\text{-}T_0$ during the forming step in zone $0.25m < y < 0.75\ m$: (a) P1 and (b) Surface

Figure 4. Deformed shapes for P1 and Surface at the end of the forming step

Results for the tempering step

     After forming, when the glass sheet was cooling, the laser was turned off. The temperature map inside the sheet at the beginning of the tempering step was the same as at the end of the forming step. Forced convection was applied with a film coefficient equal to $300\ W.m^{-1}.K^{-1}$.

     The compressive and tensile zones in the glass sheet at the end of tempering for P1 and Surface are given in Figure 5. Qualitatively, the breakdown between compressive and tensile residual stresses was similar overall for both models. They only differed in two areas. Under the laser, the layer of compressive residual stress was thinner for Surface than for P1. In contrast, it was thicker for "Surface" at the right extremity.

     The distribution of the residual stresses in the thickness on line $y = 0.1\ m$ is plotted on Figure 6. The traditional parabolic shape was obtained for both models, but with very different magnitudes. In the core, tensile stress was equal to $+13.8\ MPa$ for Surface andto $+47.7\ MPa$ for

P1. At the surface, compressive stress was equal to $-34.9\ MPa$ for Surface and $-89.1\ MPa$ for P1. Compared with the work of Gardon and Narayanaswamy[11], Surface predicts residual stress values after tempering that differs greatly from the experimental data, in contrast with P1. Once again, this can be explained by the overly diffusive Rosseland approximation leading to a smaller temperature gradient between the core and the surface during the tempering, and consequently, to smaller residual stresses.

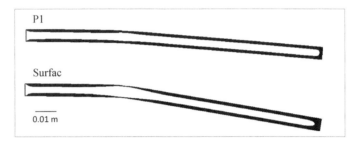

Figure 5. Compression stresses (black color) and tensile stresses (white color) in the tempered glass sheet for P1 and Surface

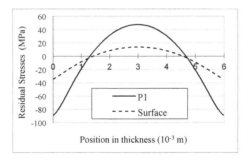

Figure 6. Distribution of residual stresses in thickness for $y = 0.1\ m$

CONCLUSION
In industrial practices, obtaining a final desired glass shape with prescribed residual stresses is one of the objectives of glass manufacturers, and radiation in the heat transfer in glass is not to be ignored.

Two-dimensional modeling of glass sheet sagging and tempering including the solution of the Radiative Transfer Equation (RTE) is developed to take surface radiation, internal radiation and external radiative sources into consideration throughout the process. The difficulty involved in solving the RTE on a non-fixed domain was overcome by adapting a numerical method implemented in commercial software.

The results of the temperature changes, the deformed shape and the residual stresses compared with those of approximate methods proved that ignoring or approximating radiative effects does not provide accurate or efficient predictive modeling.

ACKNOWLEDGMENTS

This research was supported by the International Campus on Safety and Intermodality in Transportation, the Nord/Pas-de-Calais Region, the European Community, the Regional Delegation for Research and Technology, the Ministry of Higher Education and Research and the National Center for Scientific Research. The authors gratefully acknowledge the support of these institutions. The authors are most grateful to Prof. G.W. Scherer (Department of Civil Engineering, Princeton Materials Institute, Princeton University) and to Prof. S.M. Rekhson (Cleveland State University) for assisting and guiding us in our work.

REFERENCES

1. D. Lochegnies and R.M.M. Mattheij (Eds.) in *Modelling of Glass Forming and Tempering. International Journal of Forming Processes.* Paris: Hermes Science Publications. ISBN 2-7462-0081-3, 1999.
2. D. Lochegnies and M. Cable (Eds.) in *Modelling and Control of Glass Forming and Tempering. International Journal of Forming Processes.* Paris: Hermes Science Publications. **7**(4), ISBN 2-7462-1041-X, 2004.
3. M. Sellier, M. Breitbach, H. Loch and N. Siedow, "An Iterative Algorithm for Optimal Mould Design in High-Precision Compression Moulding," in *Proceedings of The Institution of Mechanical Engineers Part B-journal of Engineering Manufacture – Proc. Inst. Mech. Eng. B-J Eng. Ma.,* **221**(1), 25-33 (2007).
4. D. Mann, R.E. Field and R. Viskanta, "Determination of Specific Heat and True Thermal Conductivity of Glass from Dynamic Temperature Data," *Heat and Mass Transfer,* **27**, 225–231 (1992).
5. S. Rosseland, "Note on the Absorption of Radiation within a Star," *M.N.R.A.S.,* **84**, 525–545 (1924).
6. J. Jiao and X. Wang, "A Numerical Simulation of Machining Glass by Dual CO_2-Laser Beams," *Optics & Laser Technology,* **40**, 297–301 (2008).
7. W. Tian and W. K. S. Chiu, "Temperature Prediction for CO_2 Laser Heating of Moving Rods," *Optics & Laser Technology,* **36**, 131–137 (2004).
8. N. Siedow, T. Grosan, D. Lochegnies, and E. Romero, "Application of a New Method for Radiative Heat Transfer to Flat Glass Tempering," *J. Amer. Ceram. Soc.,* **88**(8), 2181–2187 (2005).
9. M. F. Modest in *Radiative Heat Transfer,* Academic Press, 2003.
10. A. Farina, A. Klar, R. M. M. Mattheij, A. Mikelic, and N. Siedow, "Mathematical Models in the Manufacturing of Glass," *Lecture Notes in Mathematics,* Springer, 2011.
11. R. Gardon and O. S. Narayanaswamy, "Stress and Volume Relaxation in Annealing Flat Glass," *J. Amer. Ceram. Soc.,* **53**(7), 380-385 (1970).
12. F. T. Lentes and N. Siedow, "Three-Dimensional Radiative Heat Transfer in Glass Cooling Processes," *Glasstechnische Berichte, Glass Science and Technology,* **72**,188–196 (1999).

APPENDIX

Table 1: Elastic properties of glass

Density	$\rho = 2500. \, kg \cdot m^{-3}$
Elastic part	$E = 71. \, GPa$
	$v = 0.22$
Glass dilatation coefficient	$\alpha_g = 9.2 \; 10^{-6}$
Liquid dilatation coefficient	$\alpha_l = 18.4 \; 10^{-6}$

Table 2: Relaxation properties of glass

Shear modulus $G(t)$		Structural relaxation $M(t)$	
A_i	$\tau_i(s)$	w_i	$\tau_{si}(s)$
0.067	10.75	0.0561	$2.707 \; 10^4$
0.053	155.00	0.5074	$1.213 \; 10^5$
0.086	1406.00	0.2163	$3.297 \; 10^5$
0.230	10150.00	0.1320	$8.963 \; 10^5$
0.340	46080.00	0.0408	$2.436 \; 10^6$

Table 3: Data for shift function $\phi(\bar{r}, t)$ (8)

$T_{ref} = 746.15 \; K$
$H/R = 7.65 \; 10^4$
$x = 0.5$

Table 4: Absorption coefficients

Band number k	$\lambda_{k-1} (\mu m)$	$\lambda_k \;\; (\mu m)$	$\kappa_k (m^{-1})$
1	0.5	2.75	20
2	2.75	4.5	330
3	4.5	6.0	5000
-	6.0	∞	opaque

AN ADVANCED EXPERT CONTROL SYSTEM AND BATCH IMAGING SOFTWARE
FOR AN IMPROVED AUTOMATIC MELTER OPERATION

H.P.H. (Erik) Muijsenberg
Robert Bodi
Menno Eisenga
Glenn Neff
Glass Service USA, Inc.
Stuart, FL 34997, USA

ABSTRACT
An Advanced Expert System $ESIII^{TM}$ Control System and Batch Imaging Software from Glass Service, Inc., are utilized for an Improved Melter Operation. The Expert System $ESIII^{TM}$ offers furnace operating stability improvement that can yield improved production, and energy savings by improved and reduced operating temperatures. Additionally the Expert System $ESIII^{TM}$ can bias the energy usage in the furnace by varying the fossil fuel input to that of electric boosting to offer cost savings. The Batch Imaging Software observes the batch line within the melter and optimizes the operation in the furnace to maintain the optimal batch length control and an improved melter operation.

INTRODUCTION
One of the important aspects of any given glass furnace operation is control of the batch line within the melter. The batch line is the transition from the feed of raw materials into the furnace to the melted glass surface within the melter. Movement of this batch line, if towards the forming end of the furnace, generally entails an increase in defects within the glass production. Hence, control of the batch line is an important consideration for any furnace and control system.

The Expert System $ESIII^{TM}$, which is a model predictive control (MPC) based control system, can be coupled with some batch imaging software to help improve the furnace operation by automatic observation and control of the batch line. The Expert System $ESIII^{TM}$ which is normally controlling furnace parameters such as fuel input and profile, glass level control, furnace pressure, etc., can also be utilized as a special tool to provide for batch imaging and therefore control within the furnace.

The Expert System $ESIII^{TM}$ is replacing many of the activities of the operator. One important activity is that it can be done better by the batch imaging software to track the batch position. Then the batch imaging software can assist in the control the optimal batch line position.

An important aspect of the Expert System $ESIII^{TM}$ and batch imaging software is its uniform operation and reliability. The batch imaging software is programed to recognize a specific batch line and control it accordingly. Currently, furnace operators are controlling the batch line by intermittently observing its position and adjusting it from their intermittent observations. The Expert System $ESIII^{TM}$ and batch imaging software offers a continuous and stable operation 24 hours a day, 7 days a week. This is in contrast to a typical rotation of four (4) operating shifts of furnace operators who may have independent ideas on the best way to operate a furnace.

The Expert System $ESIII^{TM}$ control is partially replacing the operator and the PID control loops while improving the quality and saving energy upon the furnace. The Expert

System ***ESIII***™ is able to more precisely control the melter operation, batch charging and glass level, the combustion process, and excess oxygen levels.

The Expert System ***ESIII***™ can also control the working end, the forehearth and forming areas, and even an annealing lehr. Shown below in Figure 1 is a state of the art, fully automatic glass furnace control during large pull changes to the furnace. The operating data for the furnace are displayed over more than a three (3) weeks period. The furnace pull rate has changed from as high as 240 tons/day down to only 200 tons/day. Also illustrated are the corresponding crown and bottom temperatures for the furnace within the recommended operating ranges bracketed thereto, while adjustment of the fuel and boosting transformers are done by the Expert System ***ESIII***™. Hence, a more stable furnace operation is achieved.

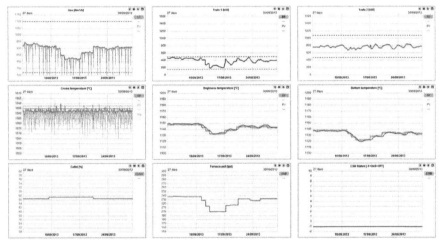

Figure 1. State-of-the-art glass furnace control data over three weeks of operation

Figure 2 shows a histogram of the bottom temperature stability on the furnace after the Expert System ***ESIII***™ was engaged for the furnace control. Previously, the bottom temperature varied between + 15°C and -15°C. The Expert System ***ESIII***™ was able to improve this to +/- 3°C. In this particular case, this was a process improvement of eighty percent (80%), which yields long-term furnace stability and improved operation.

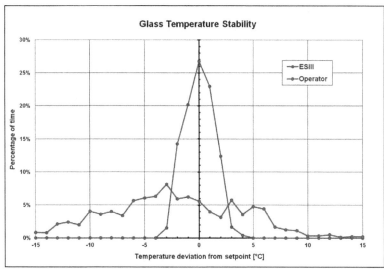

Figure 2. Histogram of the bottom temperature stability on the furnace

Figure 3 shows the various control points upon a float furnace from the perspective of that of the furnace operator. Generally, the furnace operator watches and controls the batch line by adjusting the total gas as well as observing the melter and refiner temperatures and adjusts the total gas for these temperatures as well.

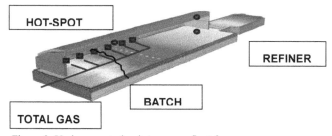

Figure 3. Various control points upon a float furnace

Figure 4 shows a furnace that is under the control of the Expert System *ESIII*™. In contrast to the operator's control, the furnace is controlled according to the entire furnace operation and as many control points that can be monitored and adjusted. The Expert System *ESIII*™ controls the furnace over multiple temperatures simultaneously and multiple individual port controls for the natural gas. Excess oxygen levels can also be maintained, and even adjusted by BTU compensation. This allows for the gas distribution to more optimally control the furnace stability. Furthermore, the batch imaging software can optimally control the batch line, while maintain a stable glass level control and throughout pull rate changes. Also noted is the refiner and canal temperature control.

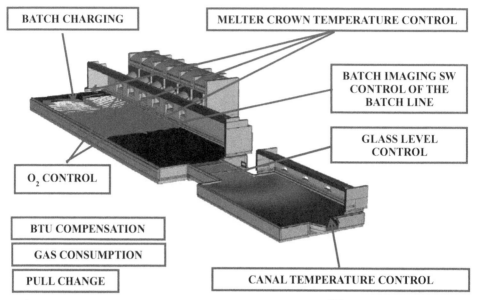

Figure 4. Furnace that is under the control of the Expert System *ESIII*™

For the Expert System *ESIII*™ to control the batch line, a camera needs to be installed at the proper position to see the batch line. Once this occurs, the Expert System *ESIII*™ can manipulate the various fuel inputs to control the batch line at the optimal position.

Shown in Figures 5 and 6 are the images seen by two (2) camera systems installed high in the back wall of a float furnace as depicted in Figure 7.

Each image shows its own side of the furnace near the batch charger, as well as an extended view down the furnace. By combining these two images through the batch imaging software, a projection of the batch piles can be made as illustrated in Figure 8. From the image, the batch piles are digitalized within the framework of the furnace geometry to determine the various batch pile positions as depicted in Figure 9. Hence, this measured data can be statistically analyzed for a given batch line within the melter.

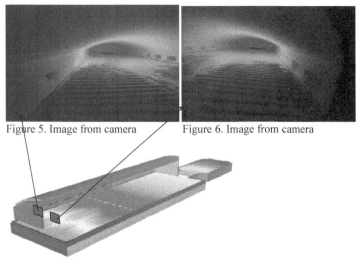

Figure 5. Image from camera   Figure 6. Image from camera

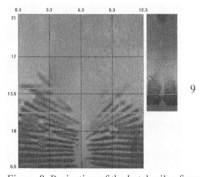

Figure 7. Positions of the cameras

Figure 8. Projection of the batch piles from images of cameras

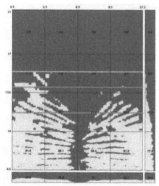

Figure 9. Digitized image of the batch piles

The furnace cameras should be installed as high as possible to obtain a bird's eye view of the melter surface. As shown in Figure 10, with only one camera installation, there is a limitation on the amount of furnace area that can be seen. Here we can see only downstream in the furnace and the back end of the furnace is not visible, nor is the right-hand side of the melter.

Figure 10. Limited view of one camera installation

If the camera is installed in the sidewall, the end of the batch line within the furnace can easily be seen. Provided there are not any furnace operational changes, Figure 11 shows the estimated view of the camera as well as an illustration of its visible area. The sidewall camera installation is acceptable, but does have its limitations on the viewable area within the melter.

Figure 11. Estimated view of the camera as well as an illustration of its visible area

The camera systems must also have a good view within the furnace. Figure 12 shows an image of buildup in front of the camera lenses, obstructing the view of some of the batch surface. Figure 13 shows an indistinct image within the furnace, as the view from the camera that is shaking slightly and/or not rigidly mounted to the furnace. It is important to the batch imaging control system to have an ideal camera position, or a camera that is rigidly mounted, with minimal or no build-up upon the lenses, as well as a good color balance.

Figure 12. Image of batch buildup   Figure 13. Indistinct image within the furnace

The image from the camera is an important issue. The camera must be able to see as much as possible within the melter, as Figure 14 only shows about fifty percent (50%) of the furnace that is available for the analyses. Figure 15 has some electrical interference.

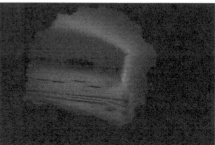

Figure 14. Limited (50%) view   Figure 15. Electrical interference

Figure 16 has some poor image quality. A clean image is what is necessary as shown in Figure 17.

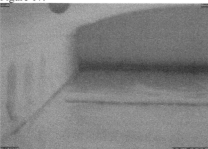

Figure 16. Poor image of quality   Figure 17. Clean image

Depending upon the size of the furnace and the view of the camera, multiple images can sometimes be used to merge into a single image showing most, if not all, of the furnace. See Figures 18 and 19 of the left and right hand images. They are then merged together to show a more complete view of the furnace in Figure 20. Only a small area in the center of the furnace near the batch chargers is not visible.

Figure 18. Left image   Figure 19. Right image

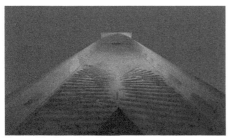

Figure 20. Complete image

Shown in Figure 21 is an orthogonal top view whereby the images from the rear of the furnace are superimposed into a top view of the melter in Figure 22. It can be seen that a much clearer vision of the batch line and position is available.

Figure 21. Orthogonal top view   Figure 22. Superimposed images in to a top view

Since a distinct and clear camera image is important to the batch imaging software, it is also important that the camera is cleaned. When the camera is removed for routine cleaning, the position of the camera will not always be returned to the same location. Hence, the batch imaging software has a special feature to automatically realign the position within the furnace.

A further consideration of the operation of the batch imaging software is a view of the batch line. This can realistically only occur when the flames are "off." In other words, the furnace reversal period is when a snapshot image of the batch line and position is taken. As shown in Figures 23, 24, and 25, you can see a series of images that are taken at various intervals that will build up the image of the batch position and the batch line. The batch imaging software takes these images of the batch during every reversal.

Figure 23. Series of images   Figure 24. Series of images   Figure 25. Series of images

Shown in Figure 26 is an illustration of a specific batch position. The batch line for this furnace can be identified at 15.4 meters in the general length as well as an indication of the furthest batch pile as noted in Figure 27. A history of the batch line can be compiled over time, generating a solid database from which to operate the furnace.

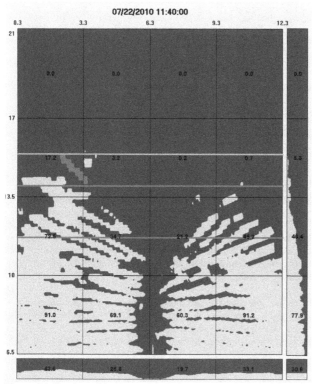

Figure 26. Batch position

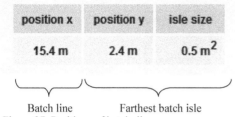

position x	position y	isle size
15.4 m	2.4 m	0.5 m^2

Batch line   Farthest batch isle

Figure 27. Positions of batchpiles

Three dimensional (3D) batch identification allows the batch thickness to also be estimated. Shown in Figure 28 is a structured grid of the melter shown by the six (6) rows and four (4) columns. The total batch coverage can thereby be estimated, and in this case the batch was shown to be 8.38 cubic meters of material. (It should be noted that this is the batch coverage is only for the surface of the glass melt, and does not take into account the un-melted batch submerged below the surface.) As shown, the batch position is at 12.2 meters. The batch thickness estimation can be utilized to increase the furnace energy as the batch thickness increases. The increased response time also helps to control the batch line.

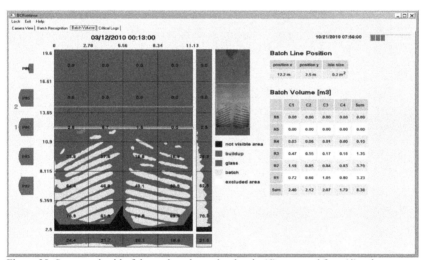

Figure 28. Structured grid of the melter shown by the six (6) rows and four (4) columns

Another illustration of the 3D batch pile coverage in a float furnace is presented in Figure 29.

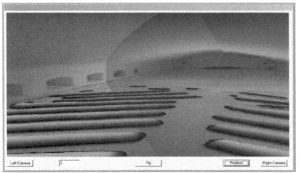

Figure 29. 3-D batch pile coverage in a float furnace

Many furnaces also utilize bubblers to help improve the mixing rate within the furnace. These bubbler positions can also be identified by the Expert System's batch imaging software and monitored accordingly. Figure 30 shows the bubblers in addition to the batch coverage.

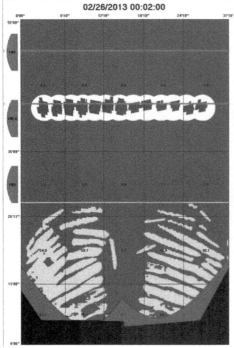

Figure 30. Bubblers and batch coverage

Many end-fired furnaces, or U-flame furnaces, will utilize only one (1) dog house. The doghouse located on one side of the furnace typically generates an uneven batch coverage within the furnace. As seen in Figure 31, the camera image can be seen over a sequence of furnace reversals. By monitoring these continuous changes to the batch pattern, the Batch Imaging Software and the Expert System *ESIII*™ can help to optimize the energy input required, including both the top-fire fossil fuel and electric boost from below.

4th past reversal   **3rd past reversal**   **2nd past reversal**   **past reversal**

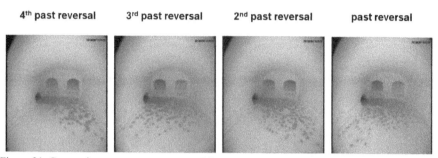

Figure 31. Camera images over a sequence of furnace reversals

An important illustration of the batch coverage is shown in Figure 32. The camera image of the furnace is shown digitalized by the batch imaging software, as well as a coverage area over the surface of the melter. This batch coverage is then tracked according to the position and coverage. The figure includes a compilation of the batch coverage, the natural gas usage, the batch charger and the batch line. By monitoring all of these factors, the furnace operation can be improved by control of the Expert System *ESIII*™.

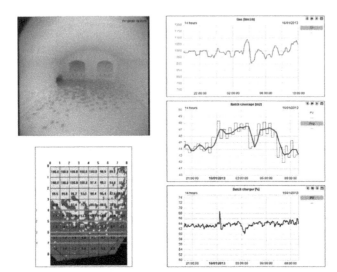

Figure 32. Camera images of batch coverage

Note the correlation of the batch position on a container glass furnace to the total gas usage and batch-charging rate.

Experience shows that the batch line position can change very fast during a furnace reversal. Therefore, it is necessary to watch the trend over several reversals to ensure optimal control. Also, a thorough and complete cleaning of the camera system should be performed periodically.

Depending upon the utilization of various camera types, multiple pictures can be combined to obtain the best analysis.

The batch imaging software provides information on the batch distribution, the batch coverage, the batch volume and relative changes to the batch line. Critical batch island positions can be noted and corrected. Upon camera cleaning, the system utilizes auto correction to reposition the imaging software for a fixed batch line position.

In summary, the batch imaging software utilized in conjunction with the Expert System *ESIII*^{TM} can keep a stable batch position, which enables a stable furnace operation. With a stable furnace operation, the furnace can be operated much more consistently, generally yielding fuel savings.

REFERENCES

[1] Josef Müller, Josef Chmelař, Robert Bódi, František Matuštík: 10 Years' Experiences with Advanced Furnace Control, Proceedings of the IX. International seminar on mathematical simulation in glass melting, Velké Karlovice 2007, p. 121

[2] Josef Müller, Robert Bódi, Josef Chmelař: How to Make Glass Furnace Control Easier: Advanced Optimal Control by Expert System *ESIII*™, Proceedings of the VIII. International seminar on mathematical simulation in glass melting, Velké Karlovice 2005, p. 200

HOW CAN PREDICTIVE STRATEGIES CONTRIBUTE TO IMPROVED POWER MANAGEMENT AND DECREASED ENERGY CONSUMPTION?

Rene Meuleman

Invensys Operations Management

Eurotherm Limited, Faraday Close, Durrington, Worthing, West Sussex B N13 3PL UK

INTRODUCTION

In co-operation with leading glass manufacturers Eurotherm have developed their next generation intelligent electrical Power Management Control System, EPower™. The EPower Power Management System is based on a Control (CPU) module which is capable of controlling up to 4 thyristor power stacks in a "PLC-like" design layout. Up to 63 of these Power Management Control Modules can be integrated together via a fast network (CAN-bus) to become a large scale intelligent power control system. This large scale Intelligent Power Control system further enhanced by Eurotherms' patented "Predictive Load Management" functionality (PLM©). Such a system is able to manage effectively the varied requirements of both small and large scale heating installations from glass bending lines, tempering furnaces and autoclaves to complete Float Glass Bath and Annealing Lehr installations. Managing these power demands to maintain an almost constant power demand and preventing load peaks and spikes.

In applications where manufactures have to control multiple loads the load management features of EPower will give these users a better control over their peak power demands. In many countries the monthly peak power demand is a critical factor in the cost that the end users have to pay for electrical energy. Our lecture will explain in detail how PLM (Predictive Load Management) works, its capabilities and the projected cost savings which can be enjoyed in large scale electrical heating system installations typified by those in the glass industry.

OBJECTIVES

In multiple full cycle firing[1] load applications, like the operation of several bending or laminating furnaces and annealing lehrs, peak power demands are likely to occur if no special measures are taken. In multiple load situations where phase angle firing is used the whole facility may suffer from a poor power factor (cos φ). Both, high peak power demands as well as poor power factor will lead to higher energy costs and increased CO_2 emissions.

Most utility companies apply a surcharge when the power factor goes below 0.9 (or 90%) or if agreed maximum power demands are exceeded. By the end of the year this can translate into thousands or even tens of thousands of dollars, depending on the size of the installation.

The "demand charge" represents the cost per kW multiplied by the greatest 15-minute demand reached in kW during the month for which the bill is rendered; however the demand is subject to power factor adjustments. Electric power suppliers reserve the right to measure such power factors at any time. Should measurements indicate that the average power factor is less than 90%; the adjusted demand will be the demand as recorded by the demand meter multiplied by 90% and divided by the percent power factor.

In addition to this energy cost penalty, these possible peak power demands result in investment in unnecessarily large (oversized) power distribution systems. Predictive Load Management (PLM) is able to eliminate those full cycle firing drawbacks resulting in more effective performance not just through simple synchronised firing or 'Load Optimisation', but through advanced load balancing and load shedding strategies provided as standard features of each Eurotherm EPower. In most applications EPower makes it possible to switch from phase angle firing into full cycle firing mode. If phase angle firing is still necessary,

133

then LTC (Load Tap Changing) strategies are able to potentially increase the power factor and help to minimise any energy cost penalties.

FUNDAMENTALS
Electrical heating systems such as glass furnace boosting, fibreglass insulation and reinforcement bushings, tin bath and annealing lehr heating systems normally use SCR[2] (semi conducting rectifier) controllers, firing in phase angle mode.

Brief overview of the Benefits and drawbacks of Phase Angle firing
Phase angle firing typically degrades the power factor while increasing harmonics and electrical noise, as shown in Figure 1.

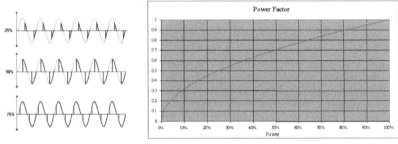

Figure 1. Phase angle firing and its effect on power factor

With phase angle firing, the power factor decreases rapidly with output power. At 50% power, the power factor is only 0.7. At 25% power, the same power factor decreases even more to 0.5. Moreover, phase angle firing creates all sorts of disturbances on the grid, such as harmonics, RFI, line losses, wasted energy (kVAr) and transformer overheating. The manufacturer will eventually be forced to increase the capacity of their equipment to compensate for these disturbances, for example by installing active or passive systems such as costly capacitors.

Conclusion :
While Phase angle firing is a simple and smooth way to control power demands with SCR's. It has two major disadvantages; poor power factor and lots of harmonics.

Power factor improvements
There are two effective methods to improve power factor in SCR driven power control systems.
- Load tap changing
- Full cycle or Burst firing

Load tap changing
Although it is outside the scope of this abstract we need to spend some time understanding how 'On Load' tap changing provides an effective way of increasing power factor of an SCR driven power system. An automatic LTC system can be used in either.
Phase Angle or Burst Firing modes of operation

By adding several taps to the transformer with a dedicated SCR for each tap, together with overlapping firing orders, such a system is capable of running at an increased power factor over a much larger range when using phase angle firing. Figure 2 shows a two tap LTC configuration running in phase angle mode and the corresponding improved power factor of such a system. By adding more taps the power factor performance increases. At the design stage it is of course necessary to have a clear understanding of the overall voltage range and the daily voltage operation range, as the best power factor can only be achieved by carefully calculating the tap-voltages.

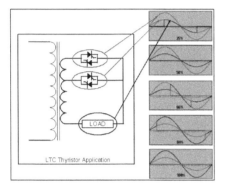 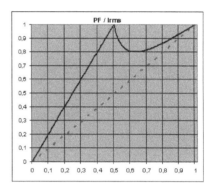

Figure 2. Two tap LTC and the effect on power factor

Full cycle firing

The easiest way to increase the power factor is to switch from phase angle to Full Cycle firing, also called Zero Cross or Burst firing. In this firing mode a modulation period is defined and inside such a modulation period the SCR is modulated with single or multiple full cycles according to the power demand. Figure 3 shows a comparison between a single SCR power system running in phase angle mode and full cycle firing mode at respectively 25%, 50% and 75% of the maximum power. Theoretically full cycle firing will result in a power factor of 1 but due to unavoidable inductive loads like transformers, wiring etc, such a system will have an overall power factor > 0.9. In fact these systems will run at the highest achievable power factor whilst avoiding the phase angle influences on the overall power factor.

Unfortunately, full cycle firing can introduce a flicker effect (main voltage variation) which in turn can affect motors and create a visual disturbance (light flicker, similar to fluorescent lighting). This effect can become more severe in multiple SCR controlled applications such as tin bath heating, annealing lehr and bushing controls when running large numbers of zones. These systems can easily contain more than 40 zones running at different power levels and variable set points. If not properly monitored, this may lead to large uncontrolled peaks of power.

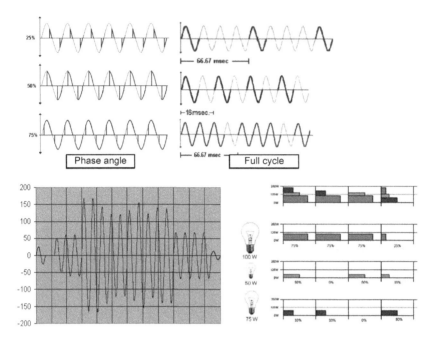

Figure 3. Comparison between a single SCR power system running in phase angle mode and full cycle firing mode

Thus, many zones randomly fired in time will increase the random peak power consumption. With two possible scenarios :

1. On new installations we therefore design in this overhead based on the maximum possible peak values, and accept the consequential additional cost of this exercise.
2. On existing installations where we may be making process improvements, changing products or increasing production capacity. This additional load may exceed the designed total power capacity of the installation resulting in possible overloads even possible black-outs.

It is however possible to address both of these scenarios firstly by ensuring that peaks of power are reduced by balancing or sharing automatically the demand across many loads to 'smooth' the average power demand on the distribution, whilst importantly still maintaining the desired power to each of these zones. Secondly by optimising and limiting the maximum allowable peak power demand of a system.

Conclusion: the power factor benefit of the full cycle firing mode provides such a big advantage that methods need to be developed to overcome the peak power related problems. One of the most sophisticated methods is Eurotherms' "Predictive Load Management".

Predictive Load Management

There are two key features of PLM which help us to address the above effectively, these are Load Balancing (or Load Sharing) and Load Shedding and are described briefly below:

Load Balancing

Overall the Load Balancing strategy is the most important part of PLM functionality, allowing us to combine multiple Burst Firing SCR's whilst maintaining a stable overall power demand.

Load Balancing is a strategy of equally distributing power of different loads to obtain an overall power consumption as stable and balanced as possible thus eliminating peaks of power. Combining multiple SCRs which are burst firing into different loads at different rates necessitates a number of sophisticated algorithms.

Each heating zone controlled by an SCR controller, is defined by an output power, cycle time and a maximum power (max capacity), which can be pictured as a rectangle. Rather than letting these rectangles pile up randomly, the PLM equipped controller uniformly distributes them thereby ensuring that at any given moment the overall power is as stable and balanced as possible. It is important to understand that the PLM function does not change the output power but rather balances and shifts the power evenly thereby eliminating any disturbance. The result is optimum load management through intelligent load balancing and load sharing, a strategy that will eliminate peaks and flicker and even cut the overall power usage, as shown in Figure 4.

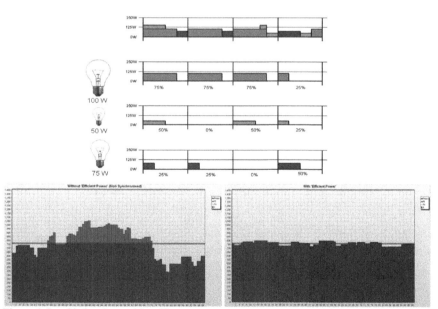

Figure 4. Load balancing and its effect on power usage

By using the PLM function, manufacturers are now able to use zero cross firing for their system without any drawbacks. Eliminating Phase Angle firing significantly improves the power factor which in turn results in substantial savings. Additionally, using energy more efficiently (i.e. substantially decreasing the reactive power (KVAR)) results in less power generated by the utility company. In fact, we should consider that consuming reactive power is in the end simply a waste of energy. While a bad power factor forces the utility company to generate this extra reactive power, it will be of absolutely no use to the end user. Besides saving costs, implementing a best practice of efficient energy consumption also results in considerably less CO_2 emissions released to the atmosphere.

Load Shedding: Demand reduction and Load Control strategy

The shedding function allows limiting and shifting the overall energy consumption all together or with fully adjustable user-defined priorities. Adjustments can be made through fieldbus communication (Profibus, DeviceNet and Ethernet) enabling dynamic adjustments in view of current PeakTime period surcharges.

Increased Quality of Main Power Supply

While an efficient load strategy can result in substantial savings and an improved environment, using synchronized SCR's will drastically increase the quality of the main supply. As described above, eliminating phase angle will remove harmful harmonics and RFI[3] generation, which for example could disturb the IT infrastructure or overheat transformers.

For customers already using full cycle firing, the PLM function can help achieve constant power balance without the flicker effect. Moreover, PLM can be of critical importance in installations where the overall installed power of the heating elements exceeds the capacity of the main transformer. In such cases non-synchronized firing may result in a total black out caused by tripped overloaded main circuit breakers. The PLM function will avoid heavy peaks of power that could overload the system by constantly monitoring and balancing the firing.

Examples

Furnace 500KW	Phase Angle		Zero Cross with EPOWER Load Balancing	
Basic Facilities Charge		$332.50		$332.50
Demand in KW	500KW		500KW	
Power Factor	0.7		>0.9	
Demand Correction KW	643KW		500KW	
Demand Charge: $9.72 per KW		$6,249.96		$4,860.00
Hour of operation	720		720	
Consumption per month (avg 50%)	180.000KWh		180.000KWh	
Energy Charge:				
First 100 kWh per Kw	$.0529/KWh	$5,290.00	$.0529/KWh	$5,290.00
Next 200 kWh per kW	$.0495/KWh	$3,960.00	$.0495/KWh	$3,960.00
Monthly Energy Cost		$15,832.46		$14,442.50
Annual Energy Cost		$189,989.52		$173,310.00
Annual Savings				$16,679.52
				-9%

Energy Costs

Load sharing, based on predictive strategies, will bring us substantial cost savings. However, proper calculation of the savings can only be made when details of the rates and conditions of the supply chain are fully known.

In addition, the possible savings in the price for the connection to the grid can be significant. Therefore Eurotherm can work with our customers at the early design stages to help them realise the full benefits prior to installing new infrastructure and negotiating energy contracts.

Many companies will have an energy manager or a specialized purchaser that knows the complexity of its company energy usage and associated costs. Both technical and purchasing personnel need to work together to find a valid technical solution to reduce energy consumption and CO_2 emissions without impact on production yield or quality.

CONCLUSION

Improving the power factor, controlling the demand charge and reducing peak consumption during ON peak times can result in substantial savings. In addition the PLM function helps to improve the quality of the main power supply and also ultimately reduce CO_2 emissions.

[1] Full cycle firing is also called burst firing mode
[2] Semi conductive rectifier
[3] Radio Frequency Interference

ACKNOWLEDGEMENTS

The author would especially like to thank Yves Level, Mikael Le Guern, Gregoire Quere, Frank Kraan, Robin Marsland and all the Eurotherm EPower team for their continuous support and contributions to this abstract.

REFERENCES

Mikaël Le Guern, Power Products Energy cost reductions through load balancing & load shedding

Yves Level, EPower Load Management Ref. 3.2.4 Load Management Option Ref. 08/07

Frank Kraan, Global Glass Cost of electrical energy Ref.10-10-2006

HOW MANY CHAMBERS ARE ENOUGH? - A FLOAT FURNACE MODELING STUDY

Matthias Lindig
Nikolaus Sorg
GmbH, Germany

OBJECTIVE OF THE STUDY
The float glass processing is predicated on the invention of the Pilkington Brother Ltd. Company dated back to 1959. The invention of this process was associated with the development of large furnace design. The design is characterized by a separation of the melting vessel in different chambers with heating and cooling sections and a complex flow pattern of the melt. The melting process and the refining and conditioning process are governed by two individual vortexes. The position of this boundary point between these two vortexes, the spring zone, is essential for the refining performance of the melter.

A modeling study was carried out in order to investigate the influence of the design modifications on the melting performances. The study has been done for a furnace with 730 metric tons per day throughput of float glass. The different number of burner ports was central in this study. Reducing the number of ports is associated with a reduction in system accessories and costs. The basin depth, the length of the refining section and the conditioning section was varied as well.

MODEL DIMENSIONS AND OPERATION BOUNDARIES
At first an existing furnace design was investigated in this study to confirm the consistency of the calculation results with the existing operation conditions. The furnace was actually designed for 800t/d (metric) throughput. The dimensions of the furnace are given in the Table 1.

Table 1. Base case furnace dimensions

base case 6port

melter area	m^2	304
refiner area	m^2	214
refiner length	m	17
conditioner area	m^2	131
melter depth	m	1,3
conditioner depth	m	1,1

Figure 1. Base case longitudinal section

The operation conditions applied in the calculation are according to the Table 2.

Table 2. Base case operation parameter

furnace pull	t/d	730
cullet	%	20
moisture	%	3
spec.pull	t/m^{2}d	2,40
Nat.gas total	Sm^3/h	4780
Combustion air	Sm^3/h	48070
conditioner cooling	Sm^3/h	6000
cooler depth	m	0,35
LCV	kJ/Sm^3	36400

The operation parameters given in Table 2 are also applied to the following model variations.

Base case modeling results
The CFD- modeling calculation was carried out for the base case using the operation parameters given above. The temperature distribution in the furnace is given in Figure 2. The checkpoints in the combustion chamber are related to direct crown temperature, those in the melter are related to direct bottom glass temperatures and the glass exit temperature.

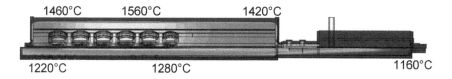

Figure 2. Base case temperature profile

The temperature distribution is in agreement with the existing furnace operation data. The length of batch blanket and the flow distribution in the combustion chamber coincide with the real furnace conditions as well.
Based on the calculation results, a particle-tracing test was carried out. The critical path is shown in Figure 3. The distance of the dots indicates the flow speed, the colour the temperature. The S-shape of the path is essential for the melting performance. Due to the back flow of the glass returning from the conditioner, the bottom glass in the melter is rising to the surface. The temperature of the melt is increasing significantly. The blister release is taking place preferentially in the denoted section.

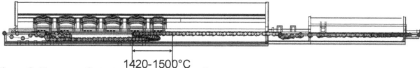

Figure 3. Base case fastest particle critical path.

Model variation results

The modifications of the furnace model are given in the Table 3.

Table 3. Furnace design parameter base case and modifications

		Standard	case study		
		6 Port	6Port	5 Port	5 Port	5 Port
melter area	m^2	304	274			
refiner area	m^2	214	184			
refiner length	m	17	15			
conditioner area	m^2	131	140			
melter depth	m	1,3	1,4	1,4	1,5	1,2
refiner depth	m	1,3	1,4	1,4	1,4	1,2
conditioner depth	m	1,1	1,1	1,1	1,1	1,1

The refiner length is shortened and the conditioner area increased. The depth of the melter was increased by 100 mm. The number of burner ports was reduced from 6 to 5 ports. The melter depth was varied as well.

The dimensions and burner port positions are shown in Figure 4. The length of the firing area in case of 6 port and 5 port is unchanged. In case of 5 ports the port width is enlarged and the distance from port to port is increased.

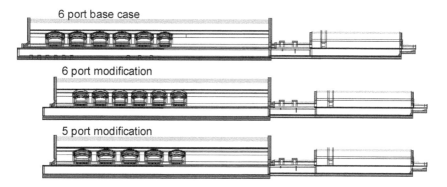

6 port base case

6 port modification

5 port modification

Figure 4. Base case and variations longitudinal section

The temperature distribution and checkpoint temperatures of the 5 and 6 port design variations are shown in Figure 5. In comparison to the base case the bottom temperatures are reduced. This is mainly due to the increased basin depth.

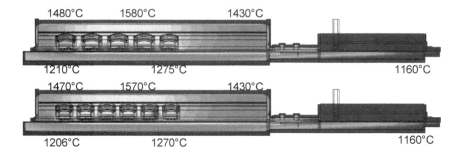

Figure 5. 5 and 6 port variation results - Temperature distribution and checkpoints.

Increasing the basin depth by an additional 100mm let the glass temperature drop an additional 10 to 15°C. The reduction in melter depth results in an increase of the glass temperature by 10°C.

For all model variations a particle tracing was performed as it was carried out for the base case.

As seen in Figures 6 and 7, the reduction in the number of burner ports did not impair the flow and refining conditions. The S-shaped critical path in both cases is very similar and does not differ from the base case.

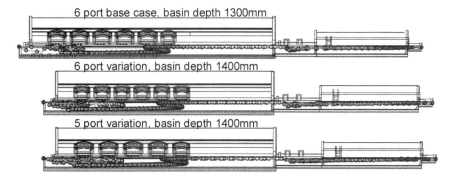

Figure 6. Base case and port number variations - Comparison of the critical path

A significant change in flow pattern appears with reduction in basin depth. The S-shaped critical path is different and the residence time and temperature of the melt in the hottest area is reduced. The melting conditions are impaired.

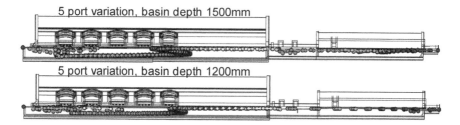

Figure 7. 5 port design with basin depth modification - Critical path

The conclusions from the comparison of the critical paths are confirmed by the calculation of the melting and fining numbers (Table 4). These numbers are used for evaluation of the particle tracing in the CFD modeling calculations. The melting number considers the residence time and temperature history of all particles passed through the model. The fining number considers only those particles with a temperature history above a fining onset temperature given by definition.

All model variations do not differ significantly in residence time and fining and melting number except the shallow basin model. The difference in fining number is significantly lower.

Table 4. Minimum residence time, melting and fining number comparison

	residence time in h	MI	FI
6 port standard	12,9	3,80E+06	2,80E+06
variations			
6 port	14	3,70E+06	2,80E+06
5 port	14	3,60E+06	2,50E+06
5 port with step	16	3,70E+06	2,60E+06
5 port shallow	12	3,50E+06	**1,80E+06**

(least 0,1% particles in target)

CONCLUSIONS
The reduction in port number from 6 to 5 does not impair the melting performance. The reduction in refiner length does not affect the refining conditions as well. The stability of the flow conditions and the sufficient residence time in the hot spot area is decisive for the melting performance. A reduction in basin depth impairs these conditions. The CFD modeling represents a suitable tool for investigating and validating different furnace design solutions. The time and temperature history is useful for estimating the refining conditions. The design evaluation regarding the number of ports and the technical decision is self-evidently based on more than modeling results. Experience and comparison with existing furnaces will also be considered. From this point of view the model study results represent a meaningful contribution to the entire assessment.

TWO-DIMENSIONAL MODELING OF THE ENTIRE GLASS SHEET FORMING PROCESS, INCLUDING RADIATIVE EFFECTS

Béchet Fabien[1,2], Siedow Norbert[3], Lochegnies Dominique[1,2]

[1] PRES Université Lille Nord de France, F-59000 Lille, France,

[2] UVHC, TEMPO, F-59313 Valenciennes, France,

[3] Fraunhofer Institute for Industrial Mathematics, Fraunhofer-Platz 1,

67663 Kaiserslautern, Germany

ABSTRACT

The entire glass sheet forming process consists of heating and forming a glass sheet and cooling and tempering it afterwards. For the first step, the glass sheet is heated using a local radiative source and deforms by sagging. In the thermo-mechanical calculations, temperature dependent glass viscosity, heat exchange with the ambient air and radiative source effects should be considered. A two-dimensional finite element model with plane deformation assumptions is developed. Using the P1-Approximation, the formulation and numerical resolution of the Radiative Transfer Equation (RTE) are performed on the glass domain as it changes over time to estimate the flux of the radiative body at each position in the glass. In the next step, the sheet is cooled. Narayanaswamy's model is used to describe the temperature dependent stress relaxation and the structural relaxation. The RTE is again solved using the P1-Approximation to consider the internal radiative effects during the cooling. There is a discussion using the P1-Approximation and comparing the results to other existing methods for the temperature changes of the glass throughout the forming process, for the deformed shape at the end of the forming step and for the residual stresses after tempering.

INTRODUCTION

Glass is a semi-transparent material with highly temperature dependent mechanical and thermal behaviors. In reality as well as in modeling, deforming glass to achieve the final desired geometry requires mastering the coupled changes of the temperature inside the glass and the product shape. Modeling a glass forming process, where large deformations occur, means computing the solution of one complex and non-linear thermo-mechanical problem. On one hand, it is necessary to determine the heat conduction of the glass taking the conditions of the boundaries between the glass and the ambient air and the glass and the forming tools into account. On the other hand, the model would also need to solve the mechanical problem with the temperature dependent viscosity of the glass and the changing contact conditions imposed by the forming tools. In the last two decades, the modeling of glass forming has been widely developed [1,2] using commercial software packages or homemade codes. For specific applications, glass must be tempered to make it more resistant and safer. During the tempering phase, the deformations are limited but there is a coupling between temperature and stress relaxation.

Besides heat conduction and heat convection, radiation plays an important role and for high temperatures, thermal radiation is the dominant heat transfer process. Very complete assessments of radiative heat transfer can be found in [3-5] and its application to the glass industry can be found in [6,7].

Since the geometry of glass changes during forming, different glass forming modeling solutions were proposed to account for radiation effects. The first and simplest one involves totally ignoring the radiation effects [8]. Another solution consists of using the Stefan-Boltzmann's law and considering only surface radiation [9]. This law is normally applied to opaque bodies, which is not the case for glass. Only a certain part of the radiant energy corresponding to the opaque spectrum of the glass is directly absorbed at the surface [10]. From a numerical point of view, the surface radiation is often approximated and taken into account by modifying the convection coefficient. This provides a linear relation instead of the T^4 non-linearity in temperature [9].

Glass is a semi-transparent material, and internal radiative effects also occur inside the glass. A widely used solution involves using equivalent conductivity (such as the active thermal conductivity method [11,12]), which is temperature dependent, or the Rosseland's approximation [13]. The Rosseland's approximation treats thermal radiation as a correction of heat conductivity, which is computed before the FEM computation using absorption coefficients. This is why it is so quick and easy to integrate into commercial software packages. It is used not only in glass forming modeling, where the domain changes overtime, but also extensively used for fixed domains with negligible deformations, such as glass tempering modeling. Originally, the method was derived in 1924 by S. Rosseland [14] to investigate stellar radiation. This is why the method is valid only for optically thick glass. Furthermore, it was shown in [15] that, for glass tempering, using the Rosseland's approximation produces vast errors in transient stress calculations.

The exact method for taking radiation in glass into account is to solve the radiative transfer equation. From a numerical point of view, this is a challenge because of the high-dimensionality and the non-linearity of RTE. A detailed discussion about different numerical methods for solving the radiative transfer equation and many more references can be found in [3] and [7]. Due to that fact that, during glass forming, one must deal with changes in the geometry of the glass plate being formed, the P1-Approximation for numerically solving the radiative transfer equation is used for the research presented here. Using the method of moments [7], one can obtain a system of two-dimensional diffusion equations instead of the high-dimensional radiative transfer equation.

In the present glass sagging modeling case, a laser is used to create local heating of the glass, as examined for the thermal computations of the cutting [16], drilling [17] or scribing [18] processes. In these studies, radiative heating was usually performed in a simple way. A first method for taking the radiative source into account is to consider it as a surface flux [16,19,20] whose intensity is directly related to the power of the radiative source. This approximation is good if the source wavelength is in the opaque zone of the glass under consideration. If this is not the case, a more precise method must be used. Another method consists of applying Beer's law, which involves nothing more than solving the one-dimensional radiative transfer equation [17,20]. In [21], Li et al. compare these two methods. The comparison reveals that both methods are very similar for optically thick glass. However, for optically thin glass, the Beer's law model provides much better results with respect to experimental data.

The research discussed here focuses on the two-dimensional (2-D) modeling of the extent of gravity sagging in a glass sheet being heated with a laser and of the tempering after sagging. This thermo-mechanical problem, for which both deformations and temperatures of the glass sheet must be computed, is a complex incremental and iterative problem that can be solved using commercial software. In the case of a deformable body discussed here, the P1-Approximation is used, for the forming and the tempering steps, to solve the RTE instead of using more simplified solutions found in the literature. An initial discussion on the temperature changes in the glass sheet during forming and on the deformed shape of the glass at the end of the forming step is proposed to compare the solution obtained by the P1-Approximation for radiation to other simplified ones. The discussion continues with the temperatures during the cooling phase and the residual stresses after tempering.

TWO-DIMENSIONAL MODELING OF GLASS SAGGING AND TEMPERING

In this paper, the glass was first exposed to a local heat using a laser source, and subsequently deformed through gravity sagging. In a second step, the deformed shape was tempered. The glass sheet was clamped on the left side ($y = 0$) and exposed to uniform convection with ambient air on all its surfaces. Considering uniform thermal conditions in the x-direction and assuming that the dimension of the sheet in the x-direction was much larger than in the (y, z)-directions, the problem can be reduced to two dimensions with generalized plane strain conditions.

Under this assumption, there is no heat transfer in the (x, z) plane but dilatation effects are allowed in the x-direction. Finally, the problem was solved on the following domain:

$$\bar{x} = (y, z), \text{and } D = \{0 \leq y \leq l, 0 \leq z \leq w,\}, D_t = D \times \{0 \leq t \leq t_{max}\}.$$

w denotes the sheet thickness, l the length (Figure 1) and t_{max} the heating duration.

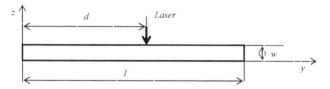

Figure 1: Description of 2-D glass sagging under radiative laser heating.

During the heating phase, the glass plate was surrounded by hot air and a laser was applied at point (d, w) with constant power. The glass is only formed due to the gravity. For the tempering step, the laser was switched off and cool air was blown all over the glass sheet.

Formulations of the gravity sagging of glass under radiative heating

The static equilibrium of the deformable glass sheet in the presence of gravitational effects without inertial effects is described using:

$$\nabla_{\bar{x}} \cdot \sigma + \rho \bar{g} = 0, \quad \bar{x} \in D, \tag{1}$$

where σ is the Cauchy stress tensor in the deformed glass sheet, ρ the density of the glass and $\bar{g}$ the gravitational force. Note that the derivative is taken with respect to position $\bar{x}$ in the actual deformed glass sheet. The boundary conditions of the glass surface ∂D are affected by a non-displacement condition imposed by clamped side ∂L_u of the glass sheet and the fact that there was no external force acting on the other glass sheet boundary $\partial D \backslash \partial L_u$. They are described using:

$$\bar{u} \cdot \bar{n} = 0, \bar{x} \in \partial L_u \qquad \text{and} \qquad \sigma \bar{n} \cdot \bar{n} = 0, \bar{x} \in \partial D \backslash \partial L_u, \tag{2}$$

where $\bar{u}$ is the displacement vector at position $\bar{x}$ and $\bar{n}$ is the normal unit vector for the glass surface.

With a very low strain rate during sagging, the glass behavior was assumed to be viscoelastic around transition temperature T_g. The elastic part is characterized by the instantaneous Young modulus E and Poisson's ratio v. At a given temperature, the glass is viscoelastic. The stress and strain tensors were split into a deviatoric tensor and a hydrostatic part using following relationships:

$$\sigma(\bar{r}, t) = s(\bar{r}, t) + \frac{trace(\sigma(\bar{r}, t))}{3} I = s(\bar{r}, t) + \frac{\sigma_h(\bar{r}, t)}{3} I, \tag{3}$$

$$\varepsilon(\bar{r},t) = e(\bar{r},t) + \frac{trace(\varepsilon(\bar{r},t))}{3}I = e(\bar{r},t) + \frac{\varepsilon_h(\bar{r},t)}{3}I, \tag{4}$$

where $e(\bar{r},t)$ is the deviatoric strain tensor, $s(\bar{r},t)$ the deviatoric stress tensor, I the unit tensor, $\varepsilon_h(\bar{r},t)$ the first strain tensor invariant and $\sigma_h(\bar{r},t)$ the first stress tensor invariant. In the following, a generalized Maxwell model was considered for the shear part. This leads to:

$$s(\bar{r},t) = \int_0^\xi G(\xi - \xi') \frac{\partial e}{\partial \xi'}(\xi')d\xi, \tag{5}$$

with shear modulus $G(t) = \sum_{i=1}^n G_i e^{-\frac{t}{\tau_i}}$. G_i is the weight at relaxation time τ_i and n the number of relaxation times used to describe the behavior of the glass (n is generally equal to 6). Bulk modulus K was used as a constant. Variable ξ was the so-called "reduced time", which was used to take temperature dependence into account through the thermo-rheological simplicity assumption. It is defined by:

$$\xi(\bar{r},t) = \int_0^t \phi[T(\bar{r},t')]dt' \tag{6}$$

where ϕ is the "shift function" defined by (8).The behavior of glass during a cooling process is very complex since structural relaxation must be taken into account. This is usually done by using the concept of fictive temperature. Roughly speaking, fictive temperature T_f represents the deviation of the structure of the glass from its equilibrium state. The fictive temperature is determined as follows:

$$T_f(\bar{r},t) = T(\bar{r},t) - \int_0^t M(\xi - \xi') \frac{dT}{d\xi'} d\xi'. \tag{7}$$

$M(t)$ is the relaxation modulus of the fictive temperature, which depends only on the material. Shift function [24] is defined by:

$$\phi(\bar{r},t) = exp\left(-\frac{H}{R_g}\left[\frac{x}{T(\bar{r},t)} + \frac{1-x}{T_f(\bar{r},t)} - \frac{1}{T_r}\right]\right). \tag{8}$$

H is an activation energy, $R_g = 8.314 \; J \cdot mol^{-1} \cdot K^{-1}$ the universal gas constant, x a material parameter and T_r a reference temperature at which $G(t)$ is measured.
Moreover, the fictive temperature T_f also contributes to thermal strain:

$$\varepsilon_{th}(\bar{r},t) = \beta_l\left(T_f(\bar{r},t) - T_0(\bar{r},t)\right) + \beta_g\left(T(\bar{r},t) - T_f(\bar{r},t)\right). \tag{9}$$

β_l and β_g are the dilatation coefficients in the liquid and solid states respectively.

Since gravity sagging will create large deformations of the glass sheet, the highly nonlinear system (1,2) must be solved to get the displacements. The high temperature dependency of the behavior of the glass (8) means that, in the modeling, we must consider the heat transfer in the glass during gravity sagging. The heat transfer in the 2-D glass sheet is described by the well-known heat transfer equation:

$$c_p \rho \frac{\partial T}{\partial t}(\bar{x}, t) = \nabla \cdot \left(k_h(\bar{x}, t)\nabla T(\bar{x}, t)\right) - \nabla \cdot \bar{q}(\bar{x}, T), \qquad (\bar{x}, t) \in D_t, \tag{10}$$

$$T(\bar{x}, 0) = T_0(\bar{x}), \qquad \bar{x} \in D. \tag{11}$$

T denotes the temperature depending on position $\bar{x}$ and time t, c_p is the specific heat capacity, and k_h the heat conductivity. $T_0(\bar{x})$ denotes the initial temperature of the glass. On the right hand side of (10), $\bar{q}(\bar{x}, T)$ denotes the radiative flux vector, which is defined as the first moment of radiative intensity $I^k(\bar{x}, \bar{\Omega}, \lambda)$ with respect to direction vector $\bar{\Omega}$ by:

$$\bar{q}(\bar{x}, T) = \int_{\lambda_{min}}^{\lambda_{max}} \int_{S^2} I(\bar{x}, \bar{\Omega}, \lambda) \, \bar{\Omega} d\Omega d\lambda. \tag{12}$$

S^2 denotes the unit sphere. At the boundary, it is proposed to describe heat flux using:

$$k_h \frac{\partial T}{\partial n}(\bar{x}, t) = \alpha\left(T(\bar{x}, t) - T^\infty(t)\right) +$$
$$\pi \epsilon \int_{opaque} \left(B(T(\bar{x}, t), \lambda) - B(T^\infty, \lambda)\right) d\lambda, \ (\bar{x}, t) \in \partial D_t. \tag{13}$$

This means that heat flux is composed of two terms. The first one represents convection with the surrounding air at a temperature T^∞. α is the convection coefficient. The second represents the difference between the radiation of the glass and the irradiation of the surroundings in the opaque wavelength region. ε denotes hemispherical emissivity and λ the wavelength in the glass, $\lambda_{min} \leq \lambda \leq \lambda_{max}$. $B(T(\bar{x}, t), \lambda)$ denotes Planck's function given as

$$B(T(\bar{x}, t), \lambda) = \frac{2hc_0^2}{n_g^2 \lambda^5 \left[e^{hc_0/n_g \lambda T(\bar{x}, t)} - 1\right]} . \tag{14}$$

k is Boltzmann's constant, h Planck's constant, c_0 the speed of light in a vacuum, and n_g the refractive index of the glass. Due to the glass sheet forming, the heat transfer problem (10-14) must be solved on a deformable body. There is coupling between the equilibrium equations (1-9) and the heat equations (10-14).

Solving the RTE during glass sagging

The glass sheet is locally heated $\left(\bar{x} \in \partial L_q\right)$ using a laser with a power denoted by $q_{laser}(t)$ acting on a surface area denoted by A_{laser}. The radiation of the hot glass in the semi-transparent wavelength region is described by the Radiative Transfer Equation (RTE) [22]. If a band model for the absorption coefficient is considered:

$$\kappa(\lambda) = \kappa_k = const., \text{ for } \quad \lambda_{k-1} \leq \lambda \leq \lambda_k, k = 1, 2, \ldots, M_k. \tag{15}$$

Then the radiative intensity must satisfy the following equation in each band k[23]:

$$\bar{\Omega} \cdot \nabla I^k(\bar{x}, \bar{\Omega}) + \kappa_k I^k(\bar{x}, \bar{\Omega}) = \kappa_k B^k\left(T(\bar{x}, t)\right), \ (\bar{x}, t) \in D_t, \tag{16}$$

with the boundary condition for $I^k(\bar{x}, \bar{\Omega})$ on domain ∂L_{qt} affected by the laser heating

$$I^k(\bar{x},\bar{\Omega}) = B^k(T^\infty) + \frac{q^k_{laser}\,(t)}{4\pi A_{laser}}, \qquad (\bar{x},t) \in \partial L_{qt}, \tag{17}$$

and the boundary condition elsewhere

$$I^k(\bar{x},\bar{\Omega}) = B^k(T^\infty), \qquad (\bar{x},t) \in \partial D_t\backslash\partial L_{qt}. \tag{18}$$

In equations (16) to (18), the following are used:

$$I^k(\bar{x},\bar{\Omega}) = \int_{\lambda_{k-1}}^{\lambda_k} I(\bar{x},\bar{\Omega},\lambda)d\lambda, \ \ B^k(T) = \int_{\lambda_{k-1}}^{\lambda_k} B(T,\lambda)d\lambda, \ \ q^k_{laser} = \int_{\lambda_{k-1}}^{\lambda_k} q^k_{laser}(\lambda)d\lambda.$$

Due to the RTE (16-18), the whole system (10-18) used to compute the temperatures in the glass sheet during glass forming is high-dimensional, non-linear, and therefore, challenging to solve numerically. This is why we use the P1-Approximation for the radiative part (16-18). Instead of (16) with boundary conditions (17) and (18), this approximation leads to

$$-\nabla \cdot \left(\frac{1}{3\kappa_k}\nabla G^k(\bar{x})\right) + \kappa_k G^k(\bar{x}) = 4\pi\kappa_k B^k\big(T(\bar{x},t)\big), \ \ (\bar{x},t) \in D_t, \tag{19}$$

$$\frac{1}{3\kappa_k}\frac{\partial G^k}{\partial n}(\bar{x}) = \frac{1}{2}\Big(G^k_a(\bar{x}) - G^k(\bar{x})\Big), \tag{20}$$

where

$$G^k_a(\bar{x}) = 4\pi B^k(T^\infty), \ \ (\bar{x},t) \in \partial D_t\backslash\partial L_{qt}, \tag{21}$$

$$G^k_a(\bar{x}) = 4\pi B^k(T^\infty) + \frac{2q^k_{laser}(t)}{A_{laser}}, \ \ (\bar{x},t) \in \partial L_{qt}. \tag{22}$$

In Equations (10-17), ∂D and ∂L_q describe the surface of the deformable glass sheet, regardless of the state of deformation being considered during gravity sagging.

The divergence of the radiative flux in (19) can then be computed using

$$\nabla \cdot \bar{q}(\bar{x},t) = \sum_k \kappa_k \left(4\pi B^k\big(T(\bar{x},t)\big) - G^k(\bar{x})\right) \tag{23}$$

The derivation of the fundamental equations and the P1-Approximation can be found in [7]. For the mathematical analysis of the coupled system (10,11,16), (19,20), (21,22), refer to [23].

To solve the RTE using P1-Approximation, one can note that (19) is very close in form to the steady-state heat transfer equation obtained by deleting the left-hand side in (10). Not only does the equation have a similar form, but the boundary conditions do as well (comparing (19) to (13)). Only term $\kappa_k G^k(\bar{x})$, present in the P1-Approximation (19) has no equivalent in the steady-state heat transfer equation (10). Consequently, the decision was made to solve the P1-Approximation using ABAQUS® finite element software because of its ability to solve the steady-state heat equation. For this reason, using a DC2D8 thermal finite element, which is an 8-node quadrangle element with bilinear interpolation in ABAQUS®, the RTE can be solved using the P1-Approximation (19) if, at each finite element level:

- $G^k(\bar{x})$ is considered as a temperature to be computed for each of the nodes of the finite element,
- $\frac{1}{3\kappa_k}$ represents material conductivity,
- the film boundary condition present in ABAQUS® is used to take (20) into account with the film coefficient equal to ½ and the film temperature equal to $G_a^k(\bar{x})$,
- the right-hand side of (20) is considered to be body flux.

By acting at each finite element level according to the aforementioned considerations to solve the RTE using P1-Approximation, the UEL (User ELement) user-subroutine in ABAQUS® was employed to add extra term $\kappa_k G^k(\bar{x})$ present in (19) and not present in the steady-state heat transfer equation (10) to the thermal stiffness matrix. With the DFLUX user-subroutine, the temperature dependent right-hand side in (19) was introduced.

The P1-Approximation (19) must be solved for each band using the aforementioned procedure in ABAQUS®. Afterwards, the divergence of the radiative flux $\nabla \cdot \bar{q}(\bar{x}, t)$ (23) is computed to determine the heat transfer in the glass sheet (10-13).

To summarize, for a given temperature map in the glass sheet (meaning for a given time t and for given temperature values at the nodes of the 2-D finite element mesh of the glass sheet), the body caused by radiation is computed as follows:

- From the temperature field and for given time t, the Planck function integrals $B^k\big(T(\bar{x}, t)\big)$ (14) to be used in (13) are computed. The surface radiation effects appearing in the second term of the boundary conditions (13) used for the heat equation (10) can be directly computed *(a first FORTRAN program was developed for this purpose)*.
- The incident radiation $G^k(\bar{x})$ (21-22) for the M_k bands is computed using functions $B^k\big(T(\bar{x}, t)\big)$ *(this is done using ABAQUS® according to the aforementioned procedure)*
- Body flux $\nabla \cdot \bar{q}(\bar{x}, t)$ is computed using results $G^k(\bar{x})$ of the M_k bands (23) *(a second FORTRAN program was developed for this purpose)*.

The method to solve the RTE with the P1-Approximation (19-22) using the ABAQUS® finite element software as described above was validated with a one-dimensional (1-D) problem by using just one band $\kappa_k = 10. \; m^{-1}$ to get an analytical solution. The 1-D problem was solved using 2-D rectangular geometry with insulation on the two horizontal boundaries and boundary conditions on the vertical edges described by (20). The analytical solution for G(x) and the ABAQUS® solution were consistent with each other (difference of less than 0.01% at each node) using a mesh of 50 uniform elements in the 1-D solution. Several κ_k values were tested. The conclusion was that the mesh must be refined near the boundaries as κ_k increases (i.e., for the more opaque bands). This is due to the appearance of boundary layers when κ_k becomes large. The coupling of the RTE with the heat transfer equation (10-13) was also validated using a 1-D thermal problem (10-13) considering a 2-band model for radiation. The value of the surface temperature was successfully compared with the solution of a 1-D finite difference program developed by Siedow et Al. [15]. In this program, the P1-Approximation was also implemented (difference less than 0.25% for the surface temperature at each time step).

MODELING RESULTS AND DISCUSSSION

In this section, different methods to take radiation into account are compared and their influences on the temperature changes during the process, on the deformed shape of the glass sheet and on the residual stresses.

Input data for the entire forming process

The geometry described in Figure 1 was considered with thickness w equal to $6. \cdot 10^{-3} m$ and length l to $150. \cdot 10^{-3} m$. The initial uniform temperature of the glass was $T_0(\bar{x}) = 873.15 \, K$. All the mechanical, thermal and radiative data are given in Appendix. The modeling is divided into two parts. During the first 25 s of forming, the laser heating occurred at a distance d (Figure 1) of $5. \cdot 10^{-3} m$. The characteristics of the laser were: width $\delta = 3. \cdot 10^{-3} m$, surface flux $\frac{q_{laser}}{A_{laser}} = 250. \, kW \cdot m^{-2}$, and wavelength $2.75 \, \mu m \leq \lambda_{laser} \leq 4.50 \, \mu m$. Note that the laser wavelength was chosen in the $k = 2$ band of the band model for the glass absorption ((15) and Table 5). Natural convection around the sheet was considered with the temperature of the surrounding air T^{∞} equal to $873.15 \, K$ and convection coefficient α equal to $20. \, W \cdot m^{-2} \cdot K^{-1}$. For the next 250 s of tempering, the laser was switched off and the glass sheet was cooled down by forced convection described by $T^{\infty} = 273.15 \, K$ and $\alpha = 300. \, W \cdot m^{-2} \cdot K^{-1}$.

Three approaches to model the heating for the forming

Three different approaches were considered in the modeling. They differ in the way the laser source, internal radiation and surface radiation are taken into account.
* Case 1, denoted "Surface": the laser heating was modeled using a surface flux q'' applied on a zone of boundary ∂L_{qt} defined by $(d - \frac{\delta}{2} < y < d + \frac{\delta}{2}, w)$ and centered on the laser entry point located at (d, w). Term q'' was added to the right-hand side of (13) only on the area concerned by the laser flux. The surface radiation was taken into account with Stefan-Boltzmann's law [9]. The radiation effects inside the glass were modeled with the Rosseland's approximation [14] using an equivalent conductivity k_e computed from the band model used and the corresponding absorption coefficients (Table 7 in Appendix). In Case 1, body flux $\nabla \cdot \bar{q}(\bar{x}, t)$ in equation (10) was not computed and taken as equal to zero.
* Case 2, denoted "Beer": the laser was modeled with Beer's law [20] considering body flux $Q(z) = \frac{q_{laser}}{A_{laser}} e^{-\kappa_2(w-z)}$ obtained by solving the 1-D RTE [20]. In the present 2-D modeling, it was assumed that the glass region concerned by the heat flux was defined as $0 < z < w$ and $-\frac{\delta}{2} < y < d + \frac{\delta}{2}$; $A^{laser} = \delta \cdot 1$ (1 is the dimension in the x −direction). Constant κ_2 is the absorption coefficient of the glass at laser wavelength λ_{laser}; κ_2 is equal to $330. \, m^{-1}$ considering Table 5 in Appendix. The body flux $Q(z)$ is added to the right-hand side of (10). In case 2, the surface radiation was also taken into account with the Stefan-Boltzmann's law and, the radiation effects were taken into consideration using the Rosseland's approximation. As in Case 1, body flux $\nabla \cdot \bar{q}(\bar{x}, t)$ in equation (10) was not computed and taken as equal to zero.
* Case 3, denoted "P1": the laser and the radiation effects inside glass are modeled with the P1-Approximation using a three-band-model for the absorption coefficient (Table 5 in Appendix). The resolution of the thermal problem and RTE were performed made using the equations [15-23].

Surface radiation was only considered in the opaque region of the spectrum using the second expression in the right-side in (13). Since the laser wavelength belonged to $[2.75\ \mu m;\ 4.50\ \mu m]$, the laser heat was also considered in equation (16) when $k = 2$. Consequently, for $k = 1$ and $k = 3$, the term q_{laser}^k vanished in (16).

Finite element mesh

The thermo-mechanical problem of the entire glass sheet forming process was incrementally and iteratively solved, using ABAQUS® finite element software. Both mechanical and thermal equations were solved using the mesh in Figure 2. The mesh is composed of 8987 nodes and 2920 CPEG8T elements with generalized plane strain conditions, biquadratic interpolation for displacements and bilinear interpolation for temperatures. Before each time step, the RTE equation was solved for each band using the P1-Approximation (18-23) and the same mesh as in Figure 2, but with DC2D8 thermal elements to solve the RTE with the modifications described in the procedure above.

Forty elements were used in the thickness with refinement near the upper and lower glass surfaces, under the laser location and on the glass edges. The mesh was refined to get a correct estimation of the temperature gradients during the heating (forming) and cooling (tempering) steps. These gradients correspond respectively to the laser heating and to the air convection cooling. Moreover, the refinement provides to get an accurate computation of the radiative energy (19-22), of the bending effects during the forming and of the residual stresses after tempering.

Figure 2 – Finite element mesh of the glass sheet in Figure 1.

Results and discussion

The first analysis concerns the forming step (25 s). Figure 3 shows the temperature changes for P1, Surface and Beer cases in three locations: in the upper surface of the glass sheet throughout the laser entry point touched by the laser (Figure 3-a), in the mid-plane of the glass sheet under the laser (Figure 3-b), and on the lower surface under the laser (Figure 3-c). These three locations are respectively denoted $A(d, w)$, $B\left(d, \frac{w}{2}\right)$ and $C(d, 0)$.

Because of the position of the laser over the sheet, the highest temperatures are obtained at Point A (Figure 3-a) regardless of the case, and the lowest temperatures are obtained at Point C (Figure 3-c).

Regardless of the point, the temperature values obtained during the forming step in the Surface case are higher than in the Beer and P1 cases. In fact, in the modeling for the Surface case, the laser energy is totally absorbed by the glass and in contrast, in the formulation of the two other models, one part passes through the glass.

At the end of the forming step, at point A (Figure 3-a) close to the laser, the temperature obtained in the P1 case is 25. K higher than in the Beer case. On the other side of the glass sheet, Point C in Figure 3-c, the temperature obtained in P1 is 20. K lower than in Beer. Inside the glass

(Points B and C), the temperatures are lower in P1 than in Surface and Beer. As found in [3,5,22], the Rosseland's approximation is too diffusive.

The difference between Beer and Surface is greater than in Tian et Li works [20]. In fact, they used a laser with a larger laser wavelength ($10.6\ \mu m$) and at this wavelength, a larger absorption coefficient for the glass ($\approx 30000.\ m^{-1}$, i.e. quite opaque glass). In consequence, less of radiation entered in the glass in their studies than in the present study.

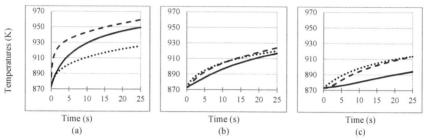

Figure 3: Temperature changes during the forming step for all cases (P1 (——), Surface (- -) and Beer (·····)) cases : (a) at Point A (d, w), (b) at Point B $\left(d, \frac{w}{2}\right)$, (c) at Point C $(d, 0)$.

At the end of the forming step, the temperature fields plotted on the initial glass geometry for the three cases are presented in Figure 4. The analysis zone is limited to $d - 3 \cdot w \le y \le d + 3 \cdot w$ (it corresponds to $32 \times 10^{-3}\ m \le y \le 68 \times 10^{-3}\ m$). Outside of this zone, the temperature remains equal to the initial temperature with no laser heating effect.

Comparing temperature fields in Figure 4, the zone concerned by the temperature changes at the end of forming is larger for Surface and Beer than for P1. In y −direction, the temperature remains equal to the initial temperature ($873.\ K$) for P1 at a distance $1.5 \cdot w$ under the laser entry point, for Surface and Beer, the distance is over $2.5 \cdot w$. There is also more homogeneous temperature distribution in z −direction for Surface and Beer compared with P1. Once again, the Rosseland's approximation is too diffusive.

Figure 5 gives the displacement in z −direction of the glass sheet middle line at the end of the forming step (25 s). With the differences in the temperature fields in Figures 3 and 4, the computed displacements on the midline are different for the P1, Surface and Beer cases.

Regardless of the case, displacement is a combination of the temperature level in the glass and of the zone concerned by the temperature change. Since glass viscosity depends on the temperature (3-8), the viscosity level is more or less reduced in a large zone of slightly varying size and glass sagging occurs. Based on the computations, one may observe that the displacements along the vertical z −axis at glass sheet edge $y = 150 \times 10^{-3}\ m$ are respectively $8.9 \times 10^{-3}\ m$, $11.6 \times 10^{-3}\ m$ and $17.2 \times 10^{-3}\ m$ for the P1, Beer and Surface cases. "This corresponds to z −displacement with thickness ratios equal to 1.48, 1.93 and 2.87. The Surface case produces the largest displacements and P1 the smallest. This can be explained by the higher heat energy absorbed by the glass in the Surface case. When comparing Beer and P1, the modeling shows that the glass region affected by the highest temperature changes is larger in case of Beer.

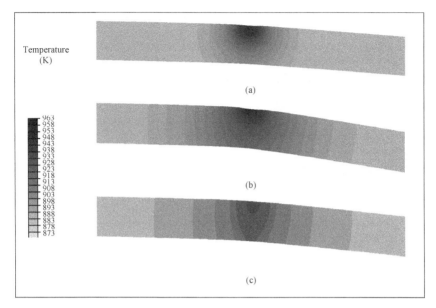

Figure 4: Temperature fields at the end of the forming step in the glass zone defined by $d - 3 \cdot w \leq y \leq d + 3 \cdot w$ (32×10^{-3} $m \leq y \leq 68 \times 10^{-3}$ m): (a) P1, (b) Surface and (c) Beer cases.

The second analysis concerns the tempering step (250 s) when the laser is shut off and the glass is cooled by forced convection. Since no laser heating is present, the term related to q_{laser} in equations (17) and (22) vanishes. The P1-Approximation (19-22) to solve the RTE remains the same as the forming step. In this tempering step, the P1-Approximation results (denoted "P1") compared with the results of the Stefan-Boltzmann's law and Rosseland's approximation are used (denoted "BS-Rosseland").

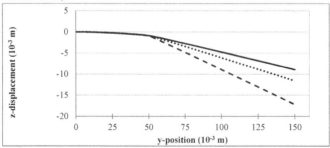

Figure 5: Representation of z-displacements along the glass sheet middle line defined by $z = \frac{w}{2}, 0 \leq y \leq l$ ($0. \leq y \leq 150. \cdot 10^{-3}$m, $z = 3 \times 10^{-3}$m) for P1 (——), Surface (- -) and Beer (·····) cases.

For the tempering, Figure 4 shows the modeling results outside of the zone affected by the laser forming, where the temperatures after the forming step remain uniform and equal to initial temperature $873.15\ K$. The analysis is made on the line defined by $y = 2 \cdot d$ and $0. \leq z \leq w$ ($y = 100$ x $10^{-3}\ m$, $0. \leq z \leq 6$ x $10^{-3}\ m$). Due to tempering conditions, there is a symmetry in the temperature distribution with respect to the glass thickness in this location.

The temperature gradient between the core and the surface of the glass sheet is one major parameter for judging the quality of the tempering [15]. Figure 6 shows the change in temperature gradient between surface and core during tempering for the SB-Rosseland and P1 cases are presented. The maximal value of the gradient is $170.K$ for P1, whereas it is only $80.K$ for SB-Rosseland. This is once again due to the overestimated diffusion of the Rosseland's approximation which leads to a smaller difference between the surface temperature and the core temperature. Consequently, the computed residual stresses in the tempered glass are different, as shown in Figure 7. The residual stress distributions along the thickness of the glass present the well-known parabolic shape with tension in the core and compression on the surface. Like the consequences of the different temperature gradients of the P1 and BS-Rosseland cases (Figure 6), the magnitudes of core and surface stresses are different. In the core, tension stress is $+13.8\ MPa$ for SB-Rosseland and, $+47.7\ MPa$ for P1. On the surface, compression stress is $-34.9\ MPa$ for SB-Rosseland and, $-89.1\ MPa$ for P1. In [24], for a convection coefficient equal to $320.\ W \cdot m^{-2} \cdot K^{-1}$ and an initial glass temperature equal to $873.15\ K$, the experiments performed by R. Gardon and O.S. Narayanaswamy showed $+50.\ MPa$ tensile stress in the core. With a convection coefficient of $300.\ W \cdot m^{-2} \cdot K^{-1}$ in the present modeling and the same initial temperature used in [24], we may conclude that:

- using the modeling of the glass tempering Stefan-Boltzmann's law and the Rosseland's approximation for surface and internal radiation in glass produces an incorrect estimation of stress,
- in contrast, using the P1-Approximation and solving the RTE produces an accurate estimation of stress.

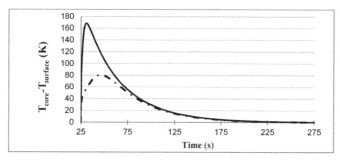

Figure 6: Temperature gradients between surface ($z = 6$ x $10^{-3}\ m$) and core ($z = 3$ x $10^{-3}\ m$) during the tempering step in location $y = 2 \cdot d = 100. \cdot 10^{-3} m$) for P1 (——) and SB-Rosseland (— ·) cases.

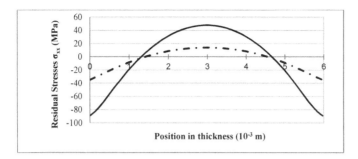

Figure 7: Distribution of residual stresses along the glass thickness in location $y = 2 \cdot d = 100 \times 10^{-3}$ m at the end of the entire glass sheet forming process for P1 (—) and SB-Rosseland (— ·) cases.

CONCLUSION

In the present study, 2-D mathematical formulations of the entire glass sheet forming process, including radiation, are proposed. In comparison with the literature, where approximated methods have been developed to account for internal and surface radiation in glass, the Radiative Transfer Equation (RTE) was solved using the P1-Approximation even through the glass domain is not a fixed domain during the forming phase. By using a band-model for the absorption coefficient of glass, the radiative body flux needed to solve the heat transfer equation was computed using ABAQUS® finite element software by developing specific user-subroutines in ABAQUS® and two complementary subroutines in FORTRAN. Comparing the results with approximated methods, it was proven that significant errors exist when the RTE is not correctly solved. These errors not only concern the temperature changes in the glass but also the deformed shape after forming and the residual stresses after tempering.

ACKNOWLEDGMENTS

This research was supported by International Campus on Safety and Intermodality in Transportation, the Nord/Pas-de-Calais Region, the European Community, the Regional Delegation for Research and Technology, the Ministry of Higher Education and Research and the National Center for Scientific Research. The authors gratefully acknowledge the support of these institutions. The authors are most grateful to Prof. G.W. Scherer (Department of Civil Engineering, Princeton Materials Institute, Princeton University) and to Prof. S.M. Rekhson (Cleveland State University) for assisting and guiding us in our work.

REFERENCES
[1] D. Lochegnies and R.M.M. Mattheij (Eds.). Modelling of Glass Forming and Tempering. International Journal of Forming Processes. Paris: Hermes Science Publications. ISBN 2-7462-0081-3, 1999.
[2] D. Lochegnies and M. Cable (Eds.). Modelling and Control of Glass Forming and Tempering. International Journal of Forming Processes. Paris: Hermes Science Publications. Vol. 7, N°4. ISBN 2-7462-1041-X, 2004.

[3] M.F. Modest. Radiative heat transfer. Academic Press, 2003.

[4] R. Siegel and J.R. Howell. Thermal radiation heat transfer. Taylor & Francis Inc., USA, 1992.

[5] R. Viskanta and E.E. Anderson. Heat transfer in semitransparent solids. In: Irvine, T.F. Jr., Harnett, J.P. (eds.) Advances in Heat Transfer,11,317–441,1975.

[6] D. Krause and H. Loch. Mathematical simulation in glass technology. Schott series on glass and glass ceramics. Springer-Verlag Berlin Heidelberg, 2002.

[7] A. Farina, A. Klar, R.M.M. Mattheij, A. Mikelic, and N. Siedow. Mathematical models in the manufacturing of glass. Lecture Notes in Mathematics, Springer, 2011.

[8] S. Gregoire, J.M.A. Cesar de Sa, P. Moreau, and D. Lochegnies. Modelling of heat transfer at glass/mould interface in press and blow forming processes. Computers and Structures, 85,15-16,1194–1205, 2007.

[9] M. Sellier, M. Breitbach, H. Loch and N. Siedow. An iterative algorithm for optimal mould design in high-precision compression moulding, Proceedings of The Institution of Mechanical Engineers Part B-journal of Engineering Manufacture – Proc. Inst. Mech. Eng. B-J Eng. Ma., 221,1,25-33, 2007.

[10] A. Sarhadi, J. Hattel, H. Hansen, C. Tutum, L. Lorenzen, and P. Skovgaard. Thermal modelling of the multi-stage heating system with variable boundary conditions in the wafer based precision glass moulding process. Journal of Materials Processing Technology, 212,8,1771-1779,2012.

[11] U. Fotheringham and F.T. Lentes. Active thermal conductivity of hot glass. Glass Science and Technology, 67,12,335-342,1994.

[12] D. Mann, R.E. Field and R. Viskanta. Determination of specific heat and true thermal conductivity of glass from dynamic temperature data. Heat and Mass Transfer, 27,225–231,1992.

[13] G. Dusserre, F. Schmidt, G. Dour, and G. Bernhart. Thermo-mechanical stresses in cast steel dies during glass pressing process. Journal of Materials Processing Technology, 162-164,484–491,2005.

[14] S. Rosseland. Note on the absorption of radiation within a star. M.N.R.A.S., 84,525–545,1924.

[15] N. Siedow, T. Grosan, D. Lochegnies, and E. Romero. Application of a new method for radiative heat transfer to flat glass tempering. Journal of the American Ceramic Society, 88,8,2181–2187,2005.

[16] J. Jiao and X. Wang. A numerical simulation of machining glass by dual CO_2-laser beams. Optics & Laser Technology,40,297–301,2008.

[17] A. Tseng, Y. Chen, C. Chao, K. Ma and T.P. Chen. Recent developments on microablation of glass materials using excimer lasers. Optics and Lasers in Engineering, 45:975–992, 2007.

[18] K. Yamamotoa, N. Hasaka, H. Morita and E. Ohmuraa. Three-dimensional thermal stress analysis on laser scribing of glass. Precision Engineering, 32,301–308,2008.

[19] A.J.C. Grellier, N.K. Zayer, and C.N. Pannell. Heat transfer modelling in CO_2 laser processing of optical fibres. Optics Communications,152,324–328,1998.

[20] W. Tian and W. K. S. Chiu. Temperature prediction for CO_2 laser heating of moving rods. Optics & Laser Technology, 36,131–137,2004.

[21] J.F. Li, L. Li and F.H. Stott. Comparison of volumetric and surface heating sources in the modeling of laser melting of ceramic materials. International Journal of Heat and Mass Transfer, 47,1159–1174,2004.

[22] F.T. Lentes and N. Siedow. Three-dimensional radiative heat transfer in glass cooling processes. Glasstechnische
Berichte, Glass Science and Technology,72,188–196,1999.

[23] R. Pinnau. Analysis of optimal boundary control for radiative heat transfer modeled by the sp1-system. Communication in Mathematical Sciences,5,4,951–969,2007.

[24] R. Gardon and O. S. Narayanaswamy, Stress and volume relaxation in annealing flat glass. Journal of the American Ceramic Society, 53,7,380-385,1970.

APPENDIX

Table 1: Elastic properties of glass

Density	$\rho = 2500.\ kg \cdot m^{-3}$
Elastic part	$E = 71.\ GPa$
	$v = 0.22$
Glass dilatation coefficient	$\alpha_g = 9.2\ 10^{-6}$
Liquid dilatation coefficient	$\alpha_l = 18.4\ 10^{-6}$

Table 2: Relaxation properties of glass (5)

Shear modulus $G(t)$		Structural relaxation $M(t)$	
G_i	$\tau_i(s)$	w_i	$\tau_{si}(s)$
0.067	10.75	0.0561	$2.707\ 10^4$
0.053	155.00	0.5074	$1.213\ 10^5$
0.086	1406.00	0.2163	$3.297\ 10^5$
0.230	10150.00	0.1320	$8.963\ 10^5$
0.340	46080.00	0.0408	$2.436\ 10^6$

Table 3: Data for shift function $\phi(\bar{r}, t)$ (8)

$T_{ref} = 746.15\ K$
$H/R = 7.65\ 10^4$
$x = 0.5$

Table 4: Thermal properties of glass

Conductivity	$k_h = 1.\ W \cdot m^{-1} \cdot K^{-1}$
Specific heat	$C_p = 1250.\ J \cdot kg^{-1} \cdot K^{-1}$
Density	$\rho = 2500.\ kg \cdot m^{-3}$

Table 5: Absorption coefficients

Band number k	$\lambda_{k-1}(\mu m)$	$\lambda_k\ (\mu m)$	$\kappa_k(m^{-1})$
1	0.50	2.75	20.
2	2.75	4.50	330.
3	4.50	6.00	5000.
-	6.00	∞	opaque

Table 6: Other radiation properties

refractive index	$n_g = 1.46$
emissivity	$\varepsilon = 0.92$

Table 7: Equivalent conductivity

$T(K)$	$k_e\ (W \cdot m^{-2} \cdot K^{-1})$
873.	7.19
923.	9.38
973.	12.04
1023.	15.19
1073.	18.86
1123.	23.07
1173.	27.83

Refractories I

HOT BOTTOM REPAIRS: GLOBAL IMPACT, PERFORMANCE CASE STUDY AND DEVELOPMENT FOR THE AMERICAS

S. Cristina Sánchez Franco
Kevin Pendleton
Dennis Cawthorn
Fosbel, Inc.
Brook Park, Ohio, USA

Bryn Snow
North American Refractories Company
West Mifflin, Pennsylvania, USA

ABSTRACT

Hot Bottom Repair (HBR) was first introduced by Fosbel, around the year 2000 and has been successfully implemented at a global level up to this date. Since then, improvements have been made to the repair processes that have resulted in an observed increase on the average repair life. An actual in-situ furnace post mortem analysis highlighting the material's performance in different areas of the furnace will be shown. In addition to presenting process application improvements, we will discuss the current state of development of the next generation HBR material, in a joint effort with North American Refractories Company.

INTRODUCTION

Hot Bottom Repair (HBR) was first introduced by Fosbel in its service portfolio by the end of 1999 start of 2000 in Europe and it was extended at a global level through the years. HBR is a patented repair process whereby a castable material is applied onto the furnace bottom under hot conditions, effectively restoring the floor thickness as is shown in Figure below, which allows the glass manufacturer to continue producing glass without having to stop the furnace for a cold repair.

| Damage Observed | During Casting | Casting Completed |

Figure 1. Illustration of HBR

According to the information at hand, Fosbel has executed 41 HBR's worldwide throughout these years. Eleven of these furnaces are still running since they were repaired (19.8 months on average, including August 2013 in the calculation). Those furnaces that underwent cold repairs after the HBR averaged 22 months of continued operation after the HBR had been performed. The longest time a furnace has operated continuously after an HBR has been 48 months and the shortest 4 months (the repair was meant to sustain the furnace for 4 months before the cold repair was scheduled).

To-date, Fosbel, through equipment design manufacturing improvements, is now able to reach areas of the furnace that are further down tank than it originally could reach, as the maximum lance length has increased from 9 m (30 feet) to 18 m (59 feet). In addition, through continuous appplication and process improvements Fosbel has been able to reduce startup defects and extend repair life from the original 12 month repair life expectancy to the current 22 month average.

Almost all repairs have been performed in container furnaces, but other market segments such as flat, fiber and tableware have benefited from this type of repair. End-fired and side-fired furnaces are the main type of furnace repaired, followed by unit melters.

HBR PERFORMANCE CASE STUDY

A post-mortem analysis of selected regions of the furnace floor was performed on a recuperative furnace that had undergone an emergency HBR. The furnace continuously ran for 25 months after the HBR and it was cooled down and repaired to increase it's melting capacity. Figure 2 shows the damaged as well as HBR cast areas. Table 1 below summarizes the furnace and HBR details,

Table 1: Case Study Furnace & HBR Information

86 m^2 (926ft^2)Recuperative Furnace	HBR Repair took 5 days
Leak occurred near furnace centerline at the bubbler line	~38,000kg (83700lb) of repair material was applied to the furnace floor
Container Glass	Furnace floor covered was approximately 65m^2 (700ft^2)
8-Year Old Furnace	Bubblers were reinstalled after the HBR
13 Bottom Bubblers	On average, ~187mm (7.3") of casted material was deposited over the bubbler strip
Underwent Color Transitions	Damaged areas had more material deposited onto them

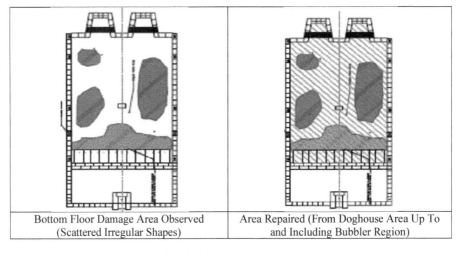

Bottom Floor Damage Area Observed (Scattered Irregular Shapes)	Area Repaired (From Doghouse Area Up To and Including Bubbler Region)

Figure 2: Damaged Area Observed and HBR Cast Area

POST MORTEM ANALYSIS AFTER 25 MONTHS OF NORMAL OPERATION

Eight (8) 3-inch (76.2mm) diameter bottom floor core drill samples were taken from selected areas of the furnace floor before it was demolished. The eight-core drill samples were chosen to represent the perceived worst areas of attack based upon the damage observed prior to repairing the furnace bottom 25 months before, when it had the emergency drain. Figure 3 below shows the location of each core drill sample, identified by number for easier reference.

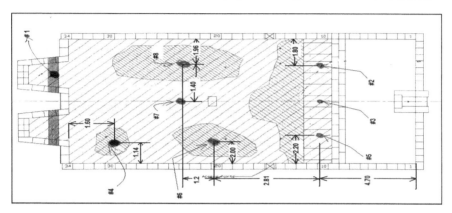

Figure 3: Core Drill Sample Location

In general, it was found that between 5 to 11 inches of applied material still remained in six (6) of the 8-core drilled samples on the furnace bottom. There was no remaining applied material on two (2) core drilled samples (Samples #2 & #3) taken at the bubbler strip region of the furnace. Table 2 below summarizes the findings for each individual core sample:

BUBBLER STRIP CORE DRILL SAMPLE DISCUSSION

Samples #2, #3 and #5 were taken from the bubbler strip region where the bubblers were re-installed by Fosbel (it is assumed they were aligned with the first original bubbler line). As such we have estimated the core sample location with the final repair configuration to be that shown in Figure 4.

Both Samples #2 & #3 had no repair material protective layer, while Sample #5 had all 11 inches of repair material. With regards to the first two samples we cannot determine the remaining bubbler block thickness as the core broke and it was not possible to drill through this block further. We believe that Sample #5 was drilled in a region between bubbler 11 and 12 and potentially was not as affected by the turbulent currents found in the immediate vicinity of a bubbler.

Table 2: Basic Furnace & HBR Information

Sample Number	Location Description	Core Photo
1	LHS Doghouse: about 5 inches of repair material still remained	
2	LHS Bubbler Strip: no repair material detected	
3	Center Line Bubbler Strip: no repair material detected	
4	RHS Blackened Floor Sample: at least 5.5" repair material remained	
5	RHS Bubbler Strip: 11" repair material remained, we believe sample was from an area between two bubblers	
6	RHS Center Region: in good shape still about 5" of repair material remained	
7	Furnace Center Line Region: all repair material (about 6.5") in excellent condition	
8	RHS Center Region: all repair material (about 11" left) in excellent condition	

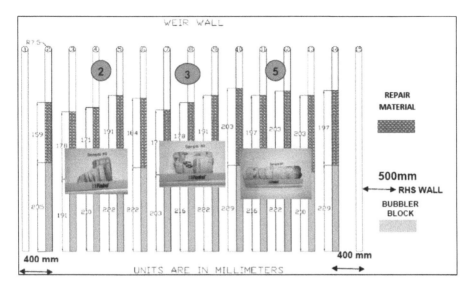

Figure 4: Bubbler Strip Detail Showing Estimated Core Sample #2, #3 & #5 Location

REMAINING CORE DRILL SAMPLES

The rest of core samples drilled had significant repair material remaining in the samples, which would indicate that these areas could potentially have survived much longer than the bubbler strip region, had the furnace not been stopped for expansion.

Core Sample #1 was of special interest because a doghouse structure iron piece had fallen onto the freshly casted material during the cullet fill following the HBR process and we wanted to evaluate its behavior on the casted mass through time. Visual inspection showed significant black discoloration of the casted mass as shown in Figure 5.

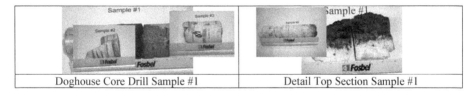

| Doghouse Core Drill Sample #1 | Detail Top Section Sample #1 |

Figure 5: LHS Doghouse Core Sample #1

Petrographic analysis of Core Sample #1 performed (1) by JTF Microscopy Services, showed that the color variations from black through pale reddish are due to the varied iron concentrations of iron-containing crystal phases observed phases (Wustite, Magnetite, Hematite, Hedenbergite and iron-rich amber glass). Paler, more beige regions below the colored ones predominantly contain the original repair material (Zirconia, Zircon, Tabular Alumina and Mullite) and bonded AZS/Glass reactant species (Albite, Nepheline and secondary Zircon). Table 3 below lists the mineral species noted above and their chemical formulae. (1)

Table 3: Crystal Phases Reported

Albite	$Na_2O-Al_2O_3-6SiO_2$
Corundum	Al_2O_3 ~ aka: "Tabular Alumina"
Cristobalite	SiO_2
Hedenbergite	$CaO-(Fe,Mg)O-2SiO_2$
Hematite	Fe_2O_3
Magnetite	$FeO-Fe_2O_3$
Mullite	$3Al_2O_3-2SiO_2$
Nepheline	$Na_2O-Al_2O_3-2SiO_2$
Wustite	FeO
Zircon	ZrO_2-SiO_2
Zirconia	ZrO_2~ aka: "Baddelyite"

HBR PERFORMANCE CASE STUDY CONCLUSIONS

All core drill samples except two (#2 & #3) had at least 5 inches or more of repair material still remaining.

Mechanical corrosion caused by the bubbling action could be the cause for the complete deterioration of the repair material protective layer found in Samples #2 & #3 at the bubbler strip region and indicating it to be the highest wear area.

Sample #5 (right hand side bubbler strip) had approximately 11 inches of repair material protective layer. The possible explanation for this is that the location where the sample was taken is calculated to have fallen between bubbler 11 and bubbler 12 in the bubbler strip.

The remaining cores (excluding those from the bubbler strip area) still had in excess of 5 inches left of repair material monolithic layer protecting the furnace for 25 months which indicates to us that the rest of furnace bottom floor could have lasted at least one additional year under the same operating conditions.

The hot bottom repair material has done a good job in encapsulating the metallic contaminants.

Petrographic analysis of Core Drill Sample #1 indicates that the repair material encapsulated and managed to contain the fallen iron in its different crystalline phase sub-products for a period of 25 months.

The encapsulation of iron by the HBR material seems to indicate improved floor protection from iron-induced floor damage, typically originating from cullet contaminated with scrap metal.

NEXT GENERATION HBR MATERIAL

Fosbel, through 41 projects of HBR, has made advances on the installation techniques and temperature profiles that have improved the performance of the repair mass while in operation. These improvements have significantly decreased the risk of stoning and extended the life of their repairs as well as allowed us to reach areas of the furnace that were previously thought to be inaccessible.

Fosbel originally used a bonded AZS monolithic material that was not developed for pumping; however, when utilizing the patented installation process and the chemical make-up of this material good results were achieved. The next phase of development was to improve on the repair material properties. With this in mind, Fosbel, in conjunction with North American Refractories Company (NARCO, an ANH Refractories Company member) is currently working on improving the flow characteristics, physical properties and corrosion resistance of the bonded AZS material used for HBR, while maintaining the same approved chemical make-up.

In ceramic materials, corrosion is by definition weight loss. This can occur by the chemical reaction between the refractory and the melt or by the erosion of refractory as the melt moves across the refractory (mechanical corrosion). Since the chemical make-up of the bonded AZS material has proven successful in the marketplace, the mechanical corrosion was the main concern during development. Mechanical corrosion is most severe when the glass is able to penetrate the refractory, so the pore size and percent porosity of the refractory should be minimized in order to resist corrosion. Another factor to consider is the viscosity of the glass, which lowers as temperature increases, allowing the melt to more easily flow into pores. Therefore, an improved material becomes critical in high temperature applications, such as in low-e glass furnace bottoms.

ANH Refractories Company research engineers applied their own self-leveling castable technology to this bonded AZS material. At ANH's product development facility, flow characteristics were quantified using an internal standardized test (2). In this vibration flow test, the frequency and the amplitude are held constant, 60HZ and 0.89kN respectively, and the sample is formed using a cone with a top diameter of 69.9 mm and a bottom diameter of 101.6mm. The measurement indicates the percent the final diameter increases compared to the original diameter of the sample after 30 seconds of vibration. In this test a higher value indicates increase flow. Table 4 shows the improvement in flow of the TEST composition over the CONTROL material.

Table 4: Flow Characteristics

Sample	Flow Value	% H_2O Added
Control	134	6.00%
Test	131	3.00%
Test	143	4.25%

The test material achieved nearly the same flow value at half the water content as the control. Less water in the system lessens the porosity of the refractory and a higher flow value results in increased densification of the monolithic material when installed through a pumping method. This is proven when comparing the physical properties of the control material versus the test material (Table 5).

In conjunction with the work done at NARCO's product development facility, samples were pumped and cast into a hot furnace at the Fosbel Brook Park, Ohio facility under the casting conditions shown in Table 6. They were soaked at 1250°C for 24 hours and then ramped down in the furnace until they reached ambient temperature. During the pumping trial, the visual observations were consistent with the findings in the laboratory.

In Figure 6, the material built up on itself to form a mound, but did not truly level out. Layers of the castable can be seen on the right side of the mound, which when pumped into the furnace would cause laminations. Figure 7 shows that the material pumped out of the hose and spread out in a flatter manner, i.e. self-leveling. In this situation, the layers are significantly smaller, so smaller lamination defects can be expected, which should result in longer refractory life.

Table 5: Physical Properties

Sample Description	Units	CONTROL	TEST	% Improvement
Casting Water Content	%	6.00%	4.25%	29%
After 1371°C Heat Treatment				
Cold Crushing Strength (ASTM C 133)	MPa	29.6	49.6	68%
Modulus of Rupture (ASTM C 133)	MPa	10.5	12.3	17%
Porosity (ASTM C 830)	%	37.8	26.4	30%
Density (ASTM C 830)	g/cm^3	2.43	2.82	16%
After 1482°C Heat Treatment				
Cold Crushing Strength (ASTM C 133)	MPa	50.7	65.9	30%
Modulus of Rupture (ASTM C 133)	MPa	12.3	15.4	25%
Porosity (ASTM C 830)	%	32.2	22.0	32%
Density (ASTM C 830)	g/cm^3	2.64	3.00	14%

Table 6: Furnace Sample Casting Parameters

Sample	Cast °T Range (°C)	% H₂O Added	Pump Pressure (Bar)	Batch Size (kg)
Control	1238 - 1256	6%	15	100
Test	1246 - 1265	4.25%	10 - 12	100

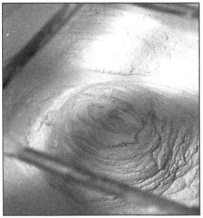 

Figure 6: Control mix flow observation   Figure 7: Test mix flow observation

Sample 1 on the left in Figure 8 is the Control and sample 2 is the Test. The plane view shows the top surface of the sample, which would be the glass contact surface in an HBR. The test sample has a much smoother top surface which would lead to less erosion from the glass melt in service.

Figure 8: Field Cast Samples, Plan view

Figure 9 shows the side view of the field cast samples; Test Sample on the left and the Control Sample on the right. In this case the test sample appeared to knit together better than the control sample which resulted in smaller and fewer laminations.

Figure 9: Field Cast Samples, Side View; Control on the right

Finally, with the aim of verifying the claim that improved physical properties, specifically porosity and density, resulted in improved corrosion resistance a static soda-lime glass corrosion test (3) was run. This test was run for 24 hours at 1427°C with the refractory samples partially submerged in a soda-lime glass melt and the results can be found in Table 7 below.

Figure 10: Glass Corrosion Fingers

Table 7. Corrosion Results

Average Loss	Control	Test
Melt Line	13.5%	9.0%
Mid-Point	1.9%	1.1%

Figure 10 shows the corrosion fingers, Control on the left and the Test on the right. In this image it is not only evident that the Test Sample had higher corrosion resistance than the Control Sample, but the appearance of the samples should also be noted. The Test finger has smaller and fewer pores than the Control finger.

Lab test results indicate that the next generation HBR material developed has shown to have improved corrosion resistance under a static soda-lime glass corrosion test, improved flow characteristics, and improved physical properties when compared to the control material that is currently used in the market.

CONCLUSIONS

HBR process has proven to be a reliable hot repair solution that allows the glass manufacturer to continue operating the furnace after the furnace floor suffers an emergency leak, or when the furnace floor is suspected to have severe damage and a potential glass leak may be imminent.

Through the years, Fosbel has executed 41 HBR worldwide, 11 of these furnaces are still in operation since they were repaired and the reamining have already undergone cold repairs. In the beginning, repair life expectancy was 12 months and currently the average repair life is 22 months.

Equipment design manufacturing improvements allows Fosbel to reach areas of the furnace that are further down tank than it originally could reach, as the maximum lance length has increased from 9m (30 feet) to 18m (59 feet).

The encapsulation of iron by the HBR material seems to indicate improved floor protection from iron-induced floor damage, typically originating from cullet contaminated with scrap metal.

Mechanical corrosion at the bubbler strip region is believed to have been the main wear mechanism controlling the repair life expectancy for the case study presented. The other furnace areas casted still had more than 5 inches of repair material protecting them, which would have allowed for a longer repair life expectancy.

A next generation bonded AZS material has been developed that improves the flow characteristics, physical properties and corrosion resistance of the material for HBR, while maintaining the same approved chemical make-up, by North American Refractories Company in conjunction with Fosbel.

ACKNOWLEDGMENTS

The authors would like to thank Mr. Terry Thomas and Ms. Karen Gallagher from Fosbel, Inc. and Mr. Nate Shadeck, and Mr. Marc Palmisiano from North American Refractories Company for their help during casting, lab analysis and proofreading.

REFERENCES

1. Fisk, J. Terry. jtf 1027 Petrographic refractory analysis of Zirmul bottom. 2009.
2. ANH Refractories. ISO 9001:2008. Vibe Flow Test. 04/07. ANH-PP-017 rev.1.
3. ISO 9001:2008. Glass Corrosion Test. 04/07. ANH-KR-024 rev.1.

PROCESS IMPROVEMENTS WITH BONDED ALUMINA CHANNELS

Elmer Sperry
Libbey Glass,
Toledo, OH 43699

Laura Lowe
North American Refractories Company
Pittsburgh, PA 15219

ABSTRACT
    Bonded Alumina Channels have recently become available to the glass industry and this paper describes the results of a trial using bonded alumina channels at Libbey Glass. The bonded alumina channels are compared to fused cast alumina channels. Forehearth construction and the effect on the process are discussed.

INTRODUCTION
    This paper is in response to the survey results from last year's conference. The participants of the conference were interested in more papers by glass makers on real experiences. This paper describes Libbey's experience with bonded alumina channels. The design of the forehearth, the effect on glass quality, the effect on energy usage and the comparison to fused alumina channels are discussed.

DISCUSSION
    In 2009 Libbey decided to experiment with a forehearth design to determine if significantly decreasing the heat loss in the forehearth would improve gob temperature homogeneity. The goal was to improve glass distribution in the forming gob and a reduction of glass distribution defects. We chose a forehearth that was ready to repair and supplied a machine that made ware susceptible to optic defects. The other favorable condition of experimenting on this forehearth was that there were several furnace and machine scheduled downtimes within the next 4 to 6 years for good inspections and opportunity to replace the channels if premature wear was noticed. This is a colorant forehearth, the color mixing section is upstream of the area that we modified and glass enters this area at elevated temperatures. Wear is generally slightly more severe where the glass enters this zone but in the area of the spout, temperatures are the same as typical forehearths. The section of this forehearth that was replaced typically has a life of over 10 years, whereas the coloring section of the forehearth has a life of 4 to 6 years depending on color production.
    The idea was that if we decreased the heat loss thru the glass contact channels of the forehearth, the temperature distribution of glass in the channel would improve and the gob would have a more even temperature distribution. The more even temperature distribution of the glass entering the bowl would improve gob temperature homogeneity and decrease glass distribution defects in the ware. We were successful at decreasing the heat loss in the forehearth. Gob temperature uniformity appeared to improve but was not measured. The forming process was not adversely affected by the change in forehearth design.
    This forehearth supplies glass to a forming machine that makes the bowls for wine glasses. The forehearth has pulls between 5 and 25 tons per day and changes pull very frequently. It operates at slightly higher temperatures than other Libbey forehearths, in the ranges from 2200 to 2100 degrees F. It is a single gob forehearth, Libbey design and Libbey designed spout and feeding equipment. This forehearth is an active colorant forehearth which is

used frequently. The area we experimented on is between the colorant stirring section and the spout. The existing fused alumina channels were replaced with bonded alumina channels in a new configuration. This zone of the forehearth was originally 26"width, with a glass depth of 6"and 8-1/4" deep fused alumina channels.

DESIGN

To improve the insulation of the forehearth we ruled out replacing the entire steel casing of the forehearth to a larger cross section because of the expense of modifying the support steel, combustion system piping and relocation of adjacent equipment. The 4 foot conditioning section casing and the spout were replaced with a larger Libbey designed casings. The insulation value was increased within the existing and new casings. The new conditioning and spout casings allowed for significant increased insulation volume. The inside width of the channels was decreased from 26" to 22". The glass depth remained the same, 6". This change did not result in the increased insulation value that was desired so the thickness of the fused alumina channels was decreased so more insulation could be placed in the forehearth casing.

Libbey switched from AZS to fused alumina channels 30 to 40 years ago as a response to cat scratch cord problems. The forehearth channels do not wear very much over the life time except at the joints, at the glass level line or at channel cracks that result from expansion stresses as a result of heating and cooling the forehearth. We considered making the channel walls and bottom thinner, reducing the 5-1/2" thickness to 3" thickness for the walls and bottom. Fused cast suppliers could not supply the channels according to our delivery requirement and the channels were expensive because the manufacturers wanted to cast thicker-walled channels and cut the sides and bottom to get the requested thickness. There were also concerns that at that thickness they would crack easily and not stay in one piece for installation. Therefore we decided to look at bonded alumina channels to solve this problem.

Libbey has had experiences with bonded channels on forehearths in the past and currently has bonded channels in operation. These are bonded AZS channels. The use of bonded AZS channels has been discontinued because occasionally these channels generate blisters, either on-going or sporadically. In another very special application bonded AZS channels were used in the condition zones of a forehearth to reduce heat loss and successfully improved the gob temperature distribution for this special, very small product. These were used for many years successfully in several forehearths but for some reason at the latest repair these channels reacted with the glass, generating blisters and had be replaced soon after installation. Bonded high alumina spouts and high alumina castables have been used in forehearths without any increase in blister defects, so we were fairly confident that the high alumina channels would perform well from a blister point of view. Bonded alumina channels have lower heat conductivity and are void free. Therefore we could decrease the thickness of the channel but still retain the required strength of the material. The corrosion rate had been reported to be similar to fused alumina channels at temperatures indicative of the application area for this trial. Figure 1 Cross section of forehearth and Figure 2 Plan view of forehearth, show the changes made to the refractory and width of the forehearth.

The vendor was hesitant to supply the 3" thick wall and bottom but agreed to a 3-1/2" thick wall and bottom. The number of channels were decreased in the assembly from 6 to 4, by increasing the length of the channels from the 24" length for the fused cast to 36-42" for the bonded alumina. The final channel design had 3-1/2" wall and bottom thickness, narrower inside width of 22" and a reduced channel height of 7" from 8-1/4".

The conditioning section of this forehearth does have stirring. The stirring is used primarily to reduce the cat scratch cord defects. The stirring system uses specially designed stirrers that have been fairly effective at diminishing cat scratch cord problems and other cord problems. The stirrers force the glass down against the top surface of the channels and mix and disrupt the flow of glass along the bottom of the channels into the spout.

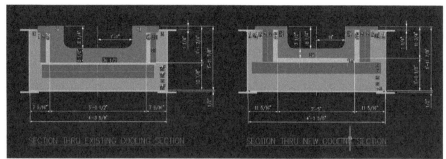

Figure 1. Comparison between the cross sections of the old design and the new design cooling section of the forehearth

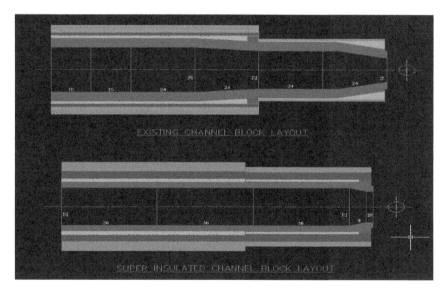

Figure 2. Plan view of the old channel layout and the new channel layout

The 3-1/2" thickness was a concern. The thin section allowed for more insulation but at the same time we assumed that the potential for glass leakage at joints and cracks would increase significantly. Glass leakage into the insulation could very quickly destroy the insulation defeating the purpose of the design change. To minimize risk, the insulation package was decreased slightly and a thicker than typical layer of high alumina ram was installed behind the channel sides and bottom, in a monolithic layer for the entire assembly. The high alumina ram layer has worked well in other forehearths preventing glass from attacking the insulation refractory. The manufacturer reports that this concept of encasing the channel with a monolithic layer has also been used historically in both soda lime and borosilicate glass with either zircon or alumina ram. This concept allows the glass maker to maximize the amount of high quality glass contact materials, including a layer with no joints to combat glass penetration. The insulation behind the glass contact and ram material was redesigned replacing insulation firebrick with ceramic fiber and micro-porous insulation.

The cost of the installation was about 15% lower primarily due to the lower volume of glass contact refractory in the assembly.

HEAT LOSS
The heat loss was calculated, estimated, thru the sides and bottom of the channel assembly.

Old Design	New Design
Side - 544 btu/hr/ft^2	Side – 208 btu/hr/ft^2
Bottom - 269 btu/hr/ft^2	Bottom – 177 btu/hr/ft^2

OPERATION
The forehearth was reheated and started up without any problems. The glass quality was normal. There was no initial blistering or increase in blistering over the campaign. There was no change in any glass defect, (seeds, blisters, stones, cord).

Temperature control was normal. Top to bottom temperature differential at the entrance to the spout improved several degrees. The firing rates decreased as a result of the improved insulation and the fuel usage was reduced. There was no significant change to the forming process.

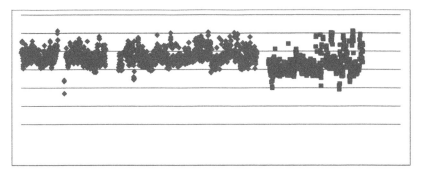

Figure 3. Graph showing the decrease in fuel usage after the design change. The fuel usage decreased by about 10% but there is a very wide range of fuel usage as a result of the pull and the glass coloring that occurs on this forehearth.

The outside temperature of the casing was noticeably cooler; the thermal images, Figures 4 and 5, indicated a temperature decrease of about 100°F between the old design, and the new design. These are different forehearths, both making the same product, one is the old design and the other the new design. The temperature line on the graph is about 8" below the burner manifold or about 12 inches below the top of the forehearthcasing.

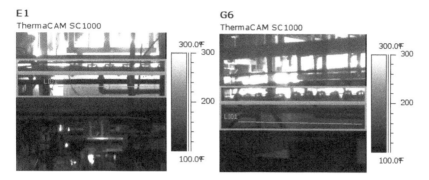

Figure 4. Thermal Photo of old design–outside of casing, Thermal Photo of new design–outside of casing

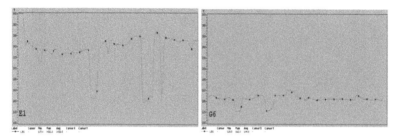

Figure 5. Temperature of the old design – outside of casing and the temperature new design – outside of casing both temperature lines approximately in the same location on the casing.

INSPECTION

At 22 months part of the forehearth was inspected at a spout change. The forehearth was not cooled down for the inspection. Cracking was noticed in the channel in the location of the stirrer. There was some pitting noticed, but not serious or particularly alarming and there were no problems that could be associated with the pitting. There was minimal joint wear which was very encouraging. The glass cut line was minimal. Thermal imaging on the outside of the casing showed no decrease in forehearth insulation value.

The crack under the stirrers was concerning. An incident several years earlier, on a different forehearth, where the conditioning zone channel had cracked and worn into the casting scar of the fused alumina channel under the stirrers had caused significant blister defects and machine downtime. It appeared that this crack could evolve into a similar condition. This is also part of a colorant forehearth so having good sound channels is important for quality. To minimize risk the material was ordered to replace the forehearth channels in the same design at the planned 48 month shut down. Figures 6-11 show various views of the forehearth and channel.

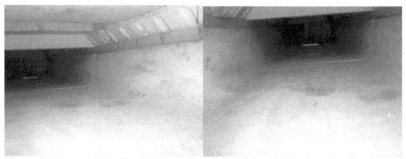

Figure 6. Photos of the forehearth at the original construction – notice some pitting on the surface.

Figure 7. Photos of the channels, covers removed. Prior to shut down this forehearth was making a blue glass and this blue glass was removed from the bottom half of the channels so the surface could be inspected.

The average wear on the surface of the channels after 48 months was about 1/8" to 3/16". The average cut at the glass level line was about 9/16" but more where the hotter glass entered the assembly and less at the spout area, the variation was about 1-1/2" at the entrance to the assembly and 3/16" adjacent to the spout.

Figure 8. Detail photos of the refractory wear points in the channel assembly - The channel wear at the glass level line was the most severe at the entrance to this section due to hot glass (2225 F to 2275 F) entering from the colorant section. The wear on the right side of the channel at the glass level was about 1-1/2" on the left side about 3/4". The channels had cracks down the center line and glass had penetrated the cracks to the ram course but not beyond that. The pitting was rather uniform with large and small holes in the surface of the refractory. Cracks down the center of the channels and up the sides did not appear to accelerate wear.

Figure 9. Photos of worn stirrer area - The channel under the stirrers was worn from about 2" to 2-1/4". The channel sides on either side of the stirrer position were worn about an additional ½" greater than upstream and downstream of that area. This wear is more than double the rate of wear seen in fused alumina channels in the same application.

Figure 10. Photos of glass penetration at cracks and joints in the channels. Glass at no place penetrated the ram course. Glass did get between the ram course and the channel refractory. This is very normal when using the monolithic ram layer under and on the sides of the fused alumina channels as well

Figure 11. Joint wear between channel blocks. The joint wear increased as the temperature increased in the forehearth. Although the hotter joint had glass penetration back to the ram, the glass did not go behind these channels on the sides or bottom of the forehearth. The joint closest to the entrance of this assemble were the most worn.

COMPARISON OF FUSED ALUMINA CHANNELS vs BONDED ALUMINA CHANNELS

The photographs in Figure 12 are an attempt to show the difference in wear rate between the bonded AZS and fused AZS channels. Temperature is critical to the wear rate of the fused alumina channels and bonded alumina channels. The fused alumina channels that operated for 6-1/2 years have less wear than those operated for 4-1/2 years, where the temperature averaged about 50 F higher.

It is not really clear in this test, how the wear compares between bonded and fused alumina channels but it indicates that wear rate accelerates faster with higher temperatures in the bonded channels than for fused cast channels. The bonded alumina channels do not wear as well as fused channels in a zone with stirrers. The wear rate at the glass line looks similar at the lower temperatures and the wear at joints and cracks possibly less especially in the lower temperature areas. The wear rate at the glass line at higher temperatures is significantly higher than fused alumina channels. The bonded channels cracked on the bottom, it could have been due to the thin section or handling, but there appears to be less cracking on the sides. The main problem is the pitting and the wear under the stirrers. The pitting appears to accelerate joint wear and surface wear.

Recently improved productions techniques have been developed that reduce surface pitting over the earlier channels that were used in our trial and this could reduce pitting long term and possibly decrease wear. The channels show in photos Figure 13 first generation channels and in Figure 14 current generation channels show significant improvement in the surface finish as compared to the channels used in our trial. The manufactures of bonded channels are continuing to make improvements, the most recent versions which we did not use in our installation, have very low surface pitting a slightly higher material density.

Photos of fused alumina channels after 6-1/2 years

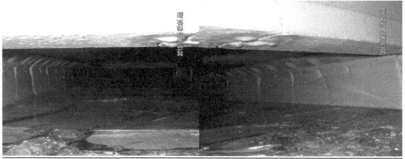

Fused alumina channels after 6.5 years – another forehearth

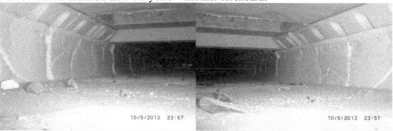

Fused alumina channels after 9.5 years – another forehearth

Fused Alumina Channels 4-1/2 years - another forehearth but hotter glass

Figure 12. 7 photos of fused alumina channels @ 6-1/2 years, 6-1/2 years, 9-1/2 years, 4-1/2 years but hotter glass.

Figure 13. First Generation surface quality

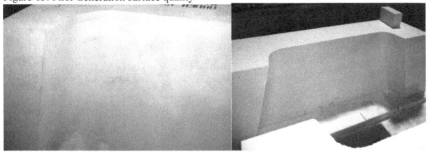

Figure 14. Current Production Technology eliminates surface pitting

CONCLUSION

This trial was a special case and not a direct replacement in the same shape as we have been using for fused alumina channels. Some of the cracking could have been less with the thicker shapes. The most disturbing problem was the observed wear rate under the stirrers which was unacceptable for this application. The pitting could have contributed to this; the pitting is also a concern for blister generation but no increased blistering was observed. The bonded alumina channels have a higher wear in higher temperature areas of the forehearth system but appear to be similar in the lower temperature areas. Bonded alumina channel may have less of a tendency to crack adjacent to the spout since it appears that bonded alumina will be less likely to crack due to the thermal shock of spout changes.

There were no other quality problems and they could have some potential benefits due to geometry flexibility over fused cast, such as reduced joints with longer channels. There could be some cost benefits.

Will we use bonded alumina channels as replacements for fused alumina? – Most likely not as a complete replacement, until the newest product technologies are fully analyzed. Although the higher wear at elevated temperatures is a concern, utilizing bonded alumina channels may be an option in our lower temperature forehearths. It is our understanding that other container manufacturers have been using this material successfully, so there indeed may be a benefit that was not realized within the scope of this trial.

BONDED REFRACTORIES FOR EXTREME CONDITIONS IN THE TOP OF REGENERATORS

Rongxing Bei, RHI AG, Industrial Division, Wiesbaden, Germany.
Klaus Santowski, RHI AG, Technology Center, Leoben, Austria.
Christian Majcenovic: RHI AG, Technology Center, Leoben, Austria.
Goetz Heilemann, RHI US Ltd., Cincinnati OH 45227
Mathew Wheeler, RHI US Ltd., Cincinnati OH 45227

ABSTRACT

The regenerator top, including the top checker layers, crown, and walls, is one of the most stressed parts of a glass melting furnace. Depending on the choice of raw materials, fuel, and operating conditions for the glass production, the chemical attack and thermal stress in the top of a regenerator can vary widely. The article reviews the historical developments behind the refractory choice for this section of the furnace. Furthermore, the results of laboratory tests examining refractory corrosion resistance to alkali, carryover, and glass melt attack are reported as well as field trials to evaluate refractory materials for regenerator top applications.

INTRODUCTION

Glass furnace regenerators have long been a topic of discussions and technical papers as glass-makers, furnace designers, and refractory producers continue to battle this harsh environment and the ever changing conditions as raw materials and new processes evolve. The regenerator top checker layers (Figure 1) are always one of the most stressed parts of a glass-melting furnace. Failure in this area can lead to clogging of the checker flues resulting in reduced efficiencies and the need for early hot repairs. The trend for energy and cost savings in the glass industry simultaneously causes more carryover (in the case of cullet preheating or batch preheating) and/or creates an aggressive atmosphere in the waste gas (in the case of alternative fuel use). Therefore, the refractories used in the regenerator top checkers face significant challenges. During prior Glass Problems Conferences, RHI has presented complete checker pack solutions as well solutions for regenerator superstructure materials. In this paper, the focus is top checker applications including history, new challenges, testing, in-service experience and refractory material choice for these extreme conditions.

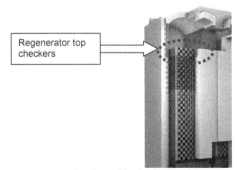

Figure 1. Regenerator chamber with checker work

HISTORICAL OVERVIEW FROM 1940's-2000

During the 1940s, magnesia bricks found their application in the glass industry, including in the top checker work layers, because fire clay bricks and silica bricks were unsatisfactory for certain highly stressed positions.[1-5] The use of magnesia refractories in the top layers enabled the furnace and checker work campaign life to be extended. The advantage of magnesia is it has a low potential to react with alkalis present in the atmosphere. Therefore, there is no so-called "alkali bursting" that is common with fireclay and silica bricks. However, because of the basic character of magnesia it reacts with SiO_2 in the carryover to form forsterite ($2MgO.SiO_2$), which causes so-called "silicate bursting", Figure 2[6,7].

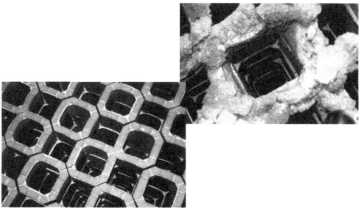

Figure 2. Silica Bursting in Top Checker Layers

Prior to the 1980's, with limited success, countermeasures to circumvent this disadvantage of silicate bursting were taken, including:

> The use of low iron and high-fired magnesia[8,9]. These materials are fired to temperatures in excess of 1800°C providing direct bonding of the MgO grains (Figure 3), thereby limiting the capacity for the grains to react with silica carryover.

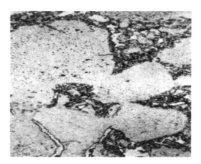

Figure 3. Direct Bonding of MgO Grains in Anker DG11

➢ Unfired, high magnesia bricks applying the theory that the unfired MgO brick is more flexible and will form forsterite when reacting with silica carryover as it is "fired" during the in the initial furnace heat up.

➢ Installing modified basic bricks, for example magnesia chromite bricks[10]. Magnesia and chromite form Mg-Cr spinel limiting the MgO free to react with silica. Chromia provides an added benefit as a highly corrosion resistant refractory component.

➢ Installing alternative refractories, such as mullite zirconium silicate bricks (e.g., ZRX)[11], which are more resistant to silica corrosion/bursting.

➢ Modifying the upper surface of the checker brick to a rounded or angled/tapered shape so the carryover cannot accumulate on the checker bricks[12].

As the above solutions were not always successful, in the 1980s, the first magnesia zircon grade was developed[13,14]. Zircon ($ZrSiO_4$) is one of the raw material components and as a result the coarse magnesia grains become protected by a forsterite ($2MgO*SiO_2$) and zirconia (ZrO_2) bonding matrix (Figure 4) that forms during the brick firing process. Since this advance, magnesia zircon bricks have become the leading choice for top layer checkers if ceramic-bonded refractories are used.

Zirconia + Forsterite

Periclase Grain

Figure 4. Magnesia-zircon microstructure.

In the late 1990s-2000, furnaces having a higher CaO content in the carryover were discovered to have problems with the magnesia zircon bricks. The CaO can react with MgO and the SiO_2 component of forsterite in the magnesia zircon bricks to form low melting Ca- and Mg-silicates (e.g., monticellite and merwinite). Therefore, since 2000, rebonded fused corundum has been used for top layer refractories. Typically, the fused corundum is inert to alkali attack if the application temperature is over 1350°C[15,16]. However, below 1350°C there is a possibility that corundum will react with alkalis from the waste gas and form ß-alumina, which causes a volume increase and results in product failure.

NEW CHALLENGES FOR TOP CHECKER LAYERS IN THE REGENERATOR

The glass industry is continuously focused on energy savings and improving environmental protection. Furthermore, there is a constant commitment to decreasing production costs. As a result multiple measures have been implemented, such as:

➢ Increasing the cullet recycling rate.

➢ Reusing dust from electric precipitators.

➢ Employing batch preheating and cullet preheating.

➢ Using alternative energy sources such as petroleum coke.

➢ Flame optimization to decrease NO_x.

All of these factors can cause more carryover (in the form of dust or even a higher proportion of fine glass cullet) or/and create an aggressive waste gas atmosphere (e.g., higher vanadium content in waste gas). This article focuses on the carryover attack as alternative fuels sources are not often utilized in the North American market and may be found in a separate article.

CORROSION TESTS

To evaluate the corrosion resistance of various refractories, two separate tests were used. First, a two-step test was performed to examine the combined attack caused by alkalis and carryover (Figure 5a). In the first step, a crucible manufactured from the refractory material to be tested was placed over a platinum crucible containing the alkali test media (i.e., Na_2CO_3 and Na_2SO_4) at 1370°C. In the second step the same refractory crucible was subsequently filled with sand, lime, and additional alkali (i.e., Na_2CO_3 and Na_2SO_4) and tested at 1470°C.

A separate test was performed to examine the interaction between the different refractory materials and glass melt (Figure 5b). In this test the refractory material crucible was filled with cullet and the test was carried out at 1470°C.

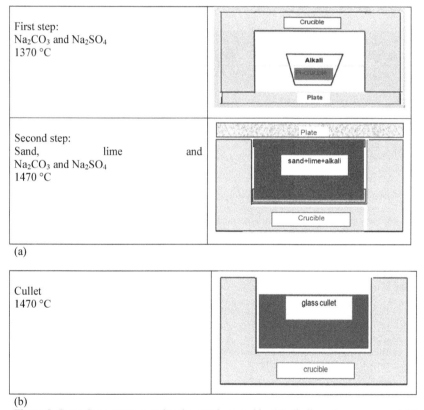

(a)

(b)
Figure 5. Corrosion tests to examine the attack caused by (a) alkalis and carryover and (b) cullet.

The test results are presented in Figure 6. The magnesia zircon brick was corroded not only by alkali and carryover (Figure 6a) but also by glass melt (Figure 6b). It is known that magnesia zircon bricks can be corroded by CaO. The CaO reacts with the SiO_2 present in the brick to form low melting Ca- and Mg-silicates[16]. Following contact with the glass melt, the magnesia grains in the magnesia zircon brick were dissolved and washed out.

As a result of alkali attack the rebonded fused corundum expanded due to the formation of ß-alumina (Figure 6c). This is why rebonded fused corundum is not recommended for low temperature applications where alkali can condense from waste gas and seriously attack the bricks. The rebonded fused corundum was also dissolved following contact with the glass melt (Figure 6d). A rebonded alpha-beta corundum brick was subsequently developed to improve performance at temperatures below 1350°C. While this composition provided an improvement over the rebonded fused corundum, some cracking was still evident as the alpha alumina transformed to beta alumina in the presence of the alkali (Figure 6e).

A zirconia mullite brand was developed for top checker layer applications using an optimized raw material concept and production parameters. The corrosion tests showed an improved performance compared to magnesia zircon and rebonded fused corundum against alkali attack and dust carryover (Figure 6f), especially against glass melt attack (Figure 6g).

The highest corrosion resistance against alkalis, carryover, and glass melt was demonstrated by the 10% chrome corundum brick and ceramic-bonded high zirconia brick. Both materials showed minimal interface corrosion (Figure 6h–6k).

REFRACTORY RECOMMENDATIONS FOR TOP CHECKER LAYERS

Table 1 summarizes the most suitable materials recommended for top checker layers depending on the specific dust situation and waste gas temperature in the regenerator top. In this table "high temperature" refers to a minimum of 1350°C whilst "lower temperature" indicates in the range of 1300–1350°C. However, it is important to stress that the temperatures are only a reference and can change due to other factors (e.g., alkali concentration in the waste gas). In summary, all factors should be considered to select an optimal solution for the top checker layers and because every situation is individual, it is important the refractory supplier determines an individual solution for the customer-specific conditions (Figure 7).

	Test with alkali and carryover	Test with cullet
Magnesia zircon RUBINAL VZ	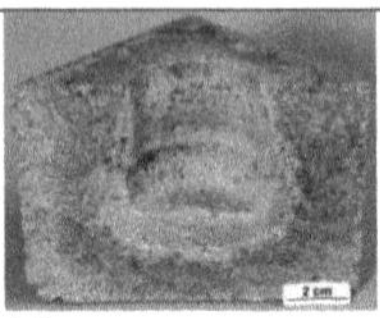 (a)	 (b)
Rebonded fused corundum DURITAL K99EXTRA	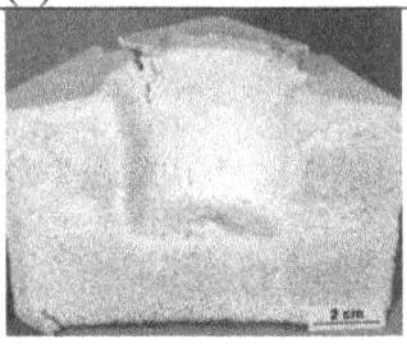 (c)	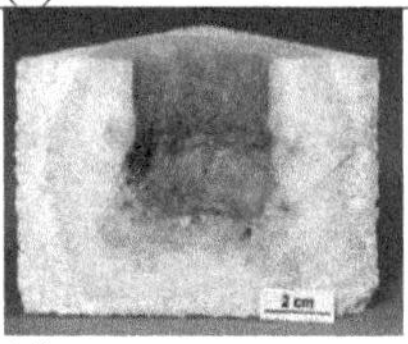 (d)
Rebonded alpha-beta corundum DURITAL K95AB	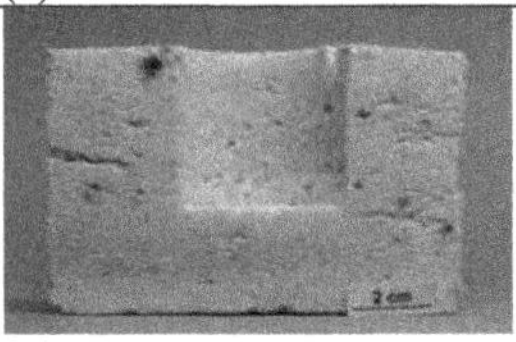 (e)	Not Available
Zirconia mullite DURITAL AZ58	 (f)	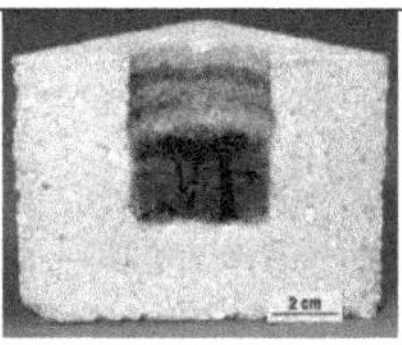 (g)
Chrome corundum DURITAL RK10	 (h)	 (i)
Bonded zirconia ZETTRAL 95CA	 (j)	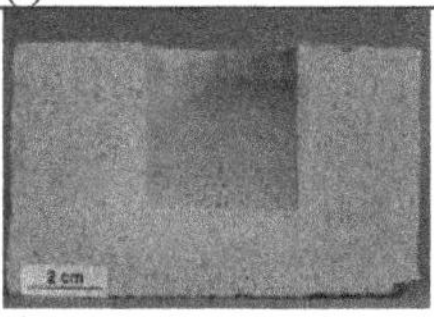 (k)

Figure 6. Crucibles after the two-step test with alkali and carryover run at 1370 and 1470°C and the test with cullet carried out at 1470°C.

Table 1. Bonded refractories recommended for the top checker layers operating under different conditions. * indicates the temperatures are only a reference and can change due to other factors (e.g., alkali concentration in the waste gas).

	Sand carryover	Sand carryover with CaO	Fine glass cullet	"Worst case" (Carryover + cullet)	
High waste gas temperature in regenerator top: > 1350°C *	Magnesia zircon	Rebonded fused corundum	Zirconia mullite	10% Chrome corundum	Bonded high ZrO2
Lower waste gas temperature in regenerator top: 1300–1350°C *	Magnesia zircon	Zirconia mullite	Zirconia mullite	10% Chrome corundum	Bonded high ZrO2

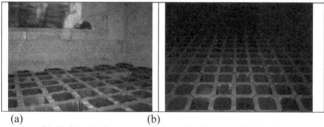

(a)   (b)

Figure 7. Examples of individual refractory solutions for the top checker layers. (a) top checker layer with chrome corundum (DURITAL RK10) for extremely high carryover and fine cullet in the waste gas and (b) top checker layer with zirconia mullite (DURITAL AZ58) for fine cullet dominating the waste gas.

CONCLUSION

Depending on the raw materials, fuel, and operating conditions in a glass melting furnace, the conditions in the regenerator top can vary quite considerably. It is important to consider all these influencing factors to optimize the refractory selection for this region. With the aid of laboratory test and field tests at the customer site, tailored solutions for each individual case can be recommended.

REFERENCES
1. E. G. Chester, "Basic Brick vs Design in Checker Brick Efficiency," *The Glass Industry*, 157–160 (1941).
2. R. G. Abbey, "Basic Regenerator Materials for Glass Furnaces," *The Glass Industry*, 268–271 (1949).
3. Unknown author, "Verwendung von Magnesitsteinen," in Glashütten. HVG-Mitteilung **627**, 1252–1255 (1953).
4. J. B. Minshall and J. C. Hicks, "Performance Report of High Magnesia Refractories in Glass Furnace Regenerators," *Ceramic Bulletin.*, 368–371 (1955).
5. M. L. Van Dreser, "Basic Refractories for the Glass Industry," *The Glass Industry*, 18–41 (1992).
6. N. Skalla, "Über Eigenschaften und Verhalten basischer Gittersteine in Glasöfen," *Glastechn. Ber.*, **33**, 169–173 (1960).
7. W. Baumgart, "Über das Verhalten eisenarmer Magnesitsteine in den Regenerativkammern von Glasschmelzwannen," *Glastechn. Ber.*, **33**, 173–182 (1960).
8. Routschka, G. and Majdic, A. Feuerfeste Baustoffe für die Glasindustrie im Speigel der Literatur. *Glas-Email-Keramo-Technik.* 1972, Books 10, 11 and 12.
9. P. Boggum in *Auswahl und Verhalten des Feuerfest-Materials in Regeneratoren*, Didier Technical Information, Wiesbaden, Germany, 1974.
10. F. Genhardt and K. Schumacher, "Verschlackungsvorgänge in den Regenerativkammern von glasschmelzanlagen zur Herstellung von Soda-Kalk-Kieselglas," Presented at XXVIIIth International Colloquium on Refractories, Aachen, Germany, 10–11, pp. 1–27, October 1985.
11. B. Schmalenbach, "Soda Lime Glass Tanks – Checker Material for Checker Chambers, Wear Mechanism and Lining Recommendation," Didier Technical Information, Wiesbaden, Germany, 1987.
12. P. Boggum, "Basiche Steine – ihre bedeutung für die Glasindustrie," *Sonderdruck aus Sprechsaal.*, 1–7 (1969).
13. B. Schmalenbach and T. Weichert, "Soda Lime Glass Tanks Checker Material for Regenerative Chambers Wear Mechanism and Lining Recommendations," *World Glass.*, **2**, 2–4 (1987).
14. T. Weichert and B. Schmalenbach, "Einsatz und Weiterentwicklung von Magnesia-Zircon-Steinen in der Glasindustrie," in *XXXVIth International Colloquium on Refractories*, Aachen, Germany, pp. 76–79, September 27-28 (1993).
15. B. Schmalenbach, DURITAL K99 Extra in Regenerator Packings, RHI Technical Information, Wiesbaden, Germany, 2004.
16. B. Schmalenbach, T. Weichert, C. Mulch, and B. Buchberger, "The Use of Ceramically Bonded Corundum and Mullite Bricks in the Superstructure and Regenerator of Glass Melting Furnaces," *RHI Bulletin,* 1, 15–19 (2006).

NEW FUSED CAST REFRACTORY FOR METAL LINE PROTECTION

Olivier Bories[a]
Isabelle Cabodi[a]
Michel Gaubil[a]
Bruno Malphettes[b]

  [a] Saint-Gobain C.R.E.E., Cavaillon, France
  [b] Saint-Gobain SEFPRO, Le Pontet, France

ABSTRACT

SEFPRO presents in this paper its new AZS for furnace over-coating at metal line. In this application, refractories are highly exposed to chemical and thermo-mechanical stresses. SEFPRO has developed a new refractory, the ER 2010 RIC® with an increased content of zirconia and an addition of yttrium, which modifies some specific properties that are critical for the target application. The main improvements are illustrated and are explained in this paper. This refractory is already industrialized, and deeply satisfies the customers.

INTRODUCTION

For furnace over-coating application, refractories are highly exposed to chemical and thermo-mechanical stresses. Indeed, the area concerned by over-coating, that is to say sidewall metal line area, is one of the most exposed areas of the furnace. The exposure is due to the very high temperatures at metal line and to the important corrosion that is observed, as illustrated in Figure 1.

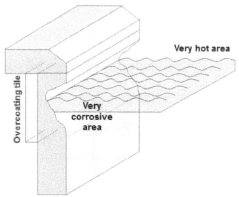

Figure 1. Metal line area is one of the most exposed to chemical and thermo-mechanical stress

The aim of side wall over-coating tiles is to protect this area from glass leakage. As a consequence, this allows the furnace to last longer.

MAIN IMPROVEMENTS OF SEFPRO REFRACTORY

SEFPRO tackles the challenge of metal line protection. Compared to the standard solution (AZS 32% zirconia) for the target application, the new ASZ brings out four main improvements:

- The corrosion resistance at metal line is improved.
- The joints are perfectly closed between the tiles.
- The thermal shock resistance is increased.
- The thermal cycling resistance is increased.

Improved Corrosion Resistance:

Dynamic corrosion tests ("Merry Go Round") have been realized. A sample of the new AZS has been compared to a sample of an AZS 32% zirconia. The samples were plunged in highly corrosive glass at high temperature. The volume difference was measured to compare both corrosion resistances. The results show an improvement of corrosion resistance of about 15%.

Lateral corrosion test has also been realized on samples of the two kinds of refractory studied above. The depth of the metal line has been measured. Results give a decrease of the metal line depth of about 30%.

Considering these results, the new AZS increases significantly the furnace lifetime since corrosion is reduced.

Perfect Joint Closure:

The joint closure is of major concern regarding the furnace lifetime. Indeed, if the joints are not perfectly closed at the operating temperature, glass can sneak into the joint and accelerate the corrosion.

A "joint closure test" has been realized on the new AZS and on an AZS 32% zirconia. The samples were heated up in a range of temperature that includes the metal line temperature and the zirconia phase transformation. As it can be seen on Figure 2, there is no joints' opening for the new AZS. As a consequence, the corrosion at the joints is highly improved as it can be seen on Figure 3.

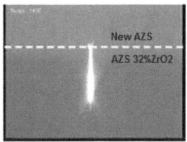

Figure 2. New AZS shows perfect joints' closure

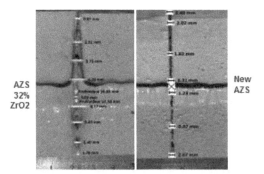

Figure 3. New AZS undergoes very little corrosion at the joints due to the perfect closure

Improved Thermal Shock Resistance:

Thermal shock resistance is also of major concern regarding the over-coating application since the installation process of over-coating tiles make the tiles undergo important thermal gradient in a short lap of time. Fig4 shows the installation process on a furnace.

Figure 4. Installation of a tile for furnace hot over-coating replacement

Hence, thermal shock tests have been realized on the new AZS and on two other AZS (at 32% and 41% of zirconia). The samples were subject to a 1250°C temperature gradient shock for five consecutive times. The material's elasticity was measured before and after the shocks, since the elasticity loss is one possible index linked to material damage. The results that are given in Figure 5 illustrate the fact that absolutely no cracks were observed on the new AZS after the test, whereas some cracks were observed on the AZS 41% of zirconia.

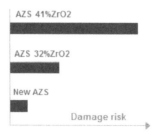

Figure 5. New AZS resists perfectly to thermal shock

Therefore, it can be stated that the new AZS is almost not sensitive to thermal shock damage.

Improved Thermal Cycling Resistance:
As the metal line is an area of the furnace that undergoes many temperature variations, thermal cycling tests have been realized.

The samples used are the new AZS and two other AZS (at 32% and 41% of zirconia). The samples were subject to cycling at the most critical temperature range. The volume variation was measured, since it is linked to material damage risk in operation. As a result, none of the samples had cracks, and the new AZS has the lowest volume variation of the three samples with 1.9%. The other samples still have very low volume variation at about 2.5%.

EXPLANATIONS OF THE IMPROVEMENTS BY THE NEW AZS CHEMICAL COMPOSITION
Compared to the standard AZS with 32% of zirconia that is eligible for the same metal line over-coating application, two main changes are made on the new AZS. The zirconia content is increased from 32% to 36%, and some yttrium is added at a 3% proportion (Table 1).

Table 1. New AZS chemical composition (in weight %)

Al_2O_3	ZrO_2	SiO_2	Y_2O_3	Others
47%	36%	13%	3%	1%

Effect of Zirconia:
The increase of the zirconia rate improves the corrosion resistance by decreasing the dissolution rate of the refractory and by limiting the Marangoni effect at the interface refractory/glass/air.

Increasing zirconia could have led to a decrease of the refractory thermal conductivity. This would have been an issue for the air blowing efficiency in the cooling operating process at the metal line. However tests have been realized to check the thermal conductivity of the new

AZS. Results, as shown in Figure 6, show that there is no decrease of the thermal conductivity compared to AZS 32% zirconia.

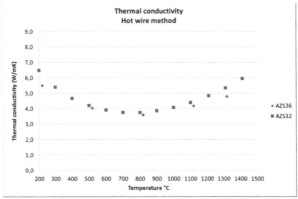

Figure 6. New AZS and AZS 32% zirconia have similar thermal conductivity

Effect of Yttrium:

On the other hand, the addition of yttrium modifies completely the thermal expansion curve of zirconia in a very favorable configuration. The graph of thermal expansion is given in fig8.

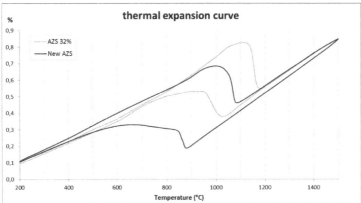

Figure 7. Effect of addition of yttrium changes on thermal expansion curve

There are two main consequences to this new thermal expansion curve:

- By decreasing the temperature and the amplitude of zirconia transformation, the temperature of perfect joint closure is reached.

- By decreasing the amplitude of the zirconia transformation and smoothing the slope of the going down expansion curve, the product is less sensitive to thermal shock and thermal cycling.

Those two consequences were illustrated in the previous section where the improvements of the new AZS were experimentally checked.

CONCLUSION

SEFPRO has developed a new AZS that is perfectly adapted to metal line over-coating application. This new AZS is named ER 2010 RIC®. The product shows four main improvements: improved corrosion resistance with no degradation of thermal conductivity, improved joint closure, improved thermal shock resistance, and improved thermal cycling resistance. These improvements are directly linked to the increase of zirconia and the addition of yttrium. The product is industrialized since 2011. It has very successful feedback in operation. ER 2010 RIC fully satisfies SEFPRO customers, as longer furnace lifetime is expected.

ANCORRO - REFINEMENT TECHNOLOGY FOR REFRACTORY IN GLASS MELT CONTACT

Rolf Weigand
Heiko Hessenkemper
Anne-Katrin Rössel
David Tritschel
Romy Kühne
Institute for Ceramic, Glass and Construction Material
TU Bergakademie Freiberg, Freiberg, Germany

ABSTRACT

The corrosion of refractory by glass melt results in many problems, which increases the production downtimes. Porous components like plunger and tube were infiltrated with glass melt after the change and create a lot of bubbles. Due to the corrosion this components have to be changed after a few months. A new technology for such components was developed at the TU Bergakademie Freiberg, Germany, which lowers the interaction between glass melt and refractory significant. The coating method of the ancorro technology for porous refractory bricks influences the glass properties in the contact area. Due to the coating an oxygen depression is generate in the pores of the bricks. These oxygen depression results in a rising of the viscosity and surface tension of the glass melt in the contact area, which slows down the attack of the refractory by glass melt. Different measurements demonstrate the influence of a reducing atmosphere to the glass properties. By realizing a lot of laboratory tests (e.g. finger tests, blistering) the interaction between glass melt and refractory can be decreased up to 90%. Also the corrosion level of fused cast bricks is realizable with porous bricks after using the coating technology. Mathematical calculations show the influence of the ancorro coating technology to the typically corrosion processes which lowers the refractory attack by glass melt. Also the nucleation, which is a problem at the orifice ring on a container glass furnace, can be prevented. During the laboratory tests only homogeneous nucleation were detected by using the coating technology on the refractory before glass contact. A lot of industrial tests on different components show the same tendency as found in the laboratory. So the ancorro coating technology was successfully implemented in industrial scale. Economical calculations demonstrate that the lower interaction between refractory and glass melt results in a bisection of the costs for production downtimes due to the changing process. In addition, prevention of the nucleation at the orifice ring allows a modification of the glass batch to a higher CaO-content. This batch modification decreases energy consumption and CO_2 emission and save costs for soda ash. The saving for a container glass producer can be up to 500,000 EUR per year and furnace.

INTRODUCTION

Rising costs for raw materials and energy making it necessary to find new ways to minimize consumption of resources in the glass industry today. Since 2008 different methods for refinement of porous refractory were investigated which lower the interaction between refractory and glass melt [1-5]. So it was found that the atmosphere in the pores of the bricks plays an essential role on the corrosion behavior of the bricks.

TECHNOLOGY

The background of the technology is to realize a "roll off effect" on the surface of the refractory as a result of a defined treatment of the porous brick. The so called oxygen depression increases the surface tension and viscosity of the glass in the boundary layer up to 30%. Due to this a layer is formed which slows down the attack of the refractory. So the interaction (e.g. corrosion, blistering) is lowered extremely. Ancorro uses over 30 different solutions for the refinement of porous refractory for glass melting to get an optimal treatment for every glass - refractory combination.

RESULTS

In 2008 the first tests were done to lower the interaction between glass melt and refractory. The technology was optimized step by step, which enables an increasing of the service life static finger tests (1450°C, 21h) up to 90% now (Figure 1). The ancorro technology can also be transferred to different glass-refractory-combinations (Figure 2), which were used in the glass industry. Industrial test confirm an increasing of the service life of porous components.

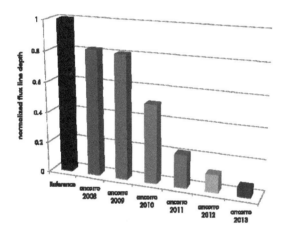

Figure 1. Lowering of the refractory corrosion since 2008

Figure 2. Effect of the ancorro-technology on different refractories, reference (left) and ancorro-treated (right): zircon-corundum, sillimanite, mullite and AZS-joint compound (from left to right)

Also typically problems during glass production can be minimized. The blistering of a porous brick decreases up to 20% to the level of fused cast material by using the ancorro treatment (Figure 3). During industrial tests the blistering of a plunger was lowered from 2 hours to 5 minutes by using the coating. This is a 95% reduction of blistering. So after the change of porous components production downtimes can be lowered which saves energy and money.

Figure 3. Samples after 12 h blistering test: porous brick, fused cast brick and porous brick refined with ancorro technology (from left to right)

After the done measurements the question arose which improvement of porous refractory is possible in comparison to fused cast material. After 96 h finger tests at 1475°C the service life of porous refractory rises to the level of fused cast bricks by using the ancorro technology (Figure 4). For the future the method should be used for porous bricks for the hot repair at the flux line in glass furnaces to increase the service life of the bricks and the whole furnace.

Figure 4. Samples after the test: porous brick, fused cast brick and porous brick refined with ancorro technology (from left to right)

Also the nucleation, which is a problem at the orifice, ring in the container glass industry can lowered (Figure 5).By using the ancorro technology the interaction and diffusion of refractory oxides into glass were lowered due to the higher viscosity and surface tension of the boundary layer. Due to this the heterogeneous nucleation of the glass can be eliminate. Industrial tests of treated orifice rings show the same effect.

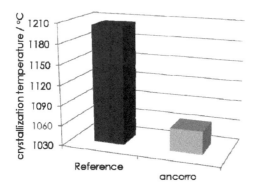

Figure 5. Lowering of the crystallization behavior of the glass in contact with refractory by using the Ancorro technology

ECONOMICAL RELEVANCE

As a result of the lower nucleation behavior of the glass in contact with refined refractory in laboratory and industrial scale a change of the glass composition is possible. The glass with the higher CaO-content shows the same chemical and mechanical properties, but the fining temperature in the furnace is 20K lower. Due to this the energy consumption, CO_2-emissions and thermal stress of the furnace is decreasing. Also expensive soda ash can be saved as a result of the increasing of the CaO-content. For a container glass furnace with a capacity of 250t/d and a furnace campaign of 10years savings of more than 500,000 EUR per year are possible.

CONCLUSION

A method for porous refractory was shown which lowers the interaction between glass melt and refractory. By using the refinement technology of ancorro porous bricks show the same properties like fused cast bricks. A decreasing of the corrosion and blistering was detected. Also the problem of heterogeneous nucleation of the glass in contact with refractory can be solved. The effect of the minimization of the interaction results in a higher viscosity and surface tension of the glass melt in the boundary layer due to the generated oxygen depression of the refinement. The technology was implemented into industrial scale, which enables savings up to 500,000 EUR per year and furnace for a container glass producer.

SPONSORING
The project ancorro is sponsored by the European Social Fund.

REFERENCES

1   H. Hessenkemper and R. Weigand, "Produktivitätserhöhung in der Glasindustrie durch Veredelung von Feuerfestmaterial," *Keramische Zeitschrift* **62**(6), 411-414 (2010).

2   R. Weigand, H. Hessenkemper, and D. Tritschel, "Ways to reduce the Interaction between Glass Melt and Refractory," *Refractories World Forum* **3**(2), 69-72 (2011).

3   R. Weigand, H. Hessenkemper, A.-K. Rössel, D. Tritschel, L. Hübner, and F. Mai, "Potential for Savings in the Container Glass Industry," *Refractories Manual* **1**, 28-32 (2012).

4   R. Weigand, H. Hessenkemper, Chr. Räbiger, A.-K Rössel, and D. Tritschel, "Bubbles as Criteria for Refractory Corrosion by Glass," *Refractories World Forum* **5**(3), 77-80 (2013).

5   R. Weigand, H. Hessenkemper, A.-KRössel, and D. Tritschel, "Influence of Atmosphere on Glass Melt Induced Refractory Corrosion," *Refractories Manual* **2**, 278-281 (2013).

Refractories II

AN UPDATE ON THE TECHNOLOGICAL EVOLUTION (OR LACK THEREOF) OF CHINESE MANUFACTURERS OF FUSED CAST REFRACTORIES AND THE VALUE vs. COST PROPOSITION

P.Carlo Ratto
Owner of fused_cast@technologist.com
San Vito al Tagliamento, PN, Italy

ABSTRACT

As a consequence of an excessive leverage on pricing that low-cost manufacturers have put into practice when trying to penetrate western markets, they have dedicated an insufficient amount of resources to improve products and develop the deeply needed level of service. Meanwhile, the cost basis of the so called "low-cost manufacturers" has been inflating (at least for the labour component), at a rate of an order magnitude higher than the cost increase of the western competitors, and only a minor fraction of this increase, until yesterday, has been transferred to Customers. This is now progressively changing at the point where China could turn into a much more "regular" supplier in pricing, and therefore it will became indispensable for them to fill up the quality/technology/service divide, to make some of the several players survive competition. In the meantime, at least for a while, the advantage/risk ratio associated to low-cost procurement could decline, and some of the western Glassmakers, which made a strategic decision to provide materials from low-cost sources, could partially review their short term strategies and/or develop specific partnership options, including technological support to the supplier. At least for a while, third parties will stay involved in the procurement chain, in order to provide services to be appended to the low-cost products, since the learning curve necessary to make these manufacturers capable to provide what western competitors are routinely offering, is still at the starting blocks.

INTRODUCTION

When, almost 20 years ago, the first Chinese fused cast refractories began to show up in the Americas and then in Europe, it was clear that there was no way for the western competitors to fight back on the commercial side.

Sale prices, sometime as low as 50% under the existing market price, where definitely lower than the western cost of manufacturing; the speed of penetration of these products in the west, however was not as high as one could imagine, thanks to very ineffective distribution channels, some evidently low qualitative aspect (mostly on the level of finishing, dimensional tolerances) and the conservative approach of glassmakers.

While western competitors had time to deploy technical and commercial strategies aimed to provide better services and technical support so as to increase customers' fidelization, most low-cost manufacturers (basically from China) did the only thing they could have done to damage themselves: they did not invest in revamping technology, increase efficiency, improve quality nor invested in developing pre- and post-sale services. They simply put all their eggs in the basket of price leverage, as shown in Figure 1 and this brute-force approach turned out to be the best mid-long term help to their competitors.

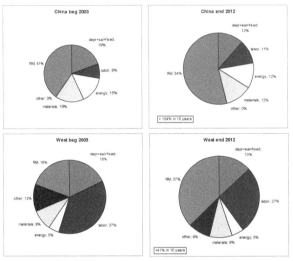

Figure 1. Price leverage of China vs. West

THE INCONTROLLABLE MACRO-ECONOMICAL FACTORS

Macro-economic factors, well beyond the producers' possibility of control (but not totally unpredictable) caused a deep transformation in the fused-cast costing structure, and this change was not symmetrical for Chinese and Western players.

While the outstanding price increase of some critical raw material (zircon sand, zirconia) hit equally both sides in absolute terms (not in relative!) as shown in Figure 2, the labor cost, that is the kernel of the Chinese competitiveness, did inflate with a yearly two-digit average rate in the recent ten years of Chinese history as shown in Figure 3. This was consequential to the necessity of expanding the huge domestic Chinese market to compensate the drop of export coming from the first globalized financial/economical crisis. Or if you like, as the relentless consequence of a globalized economy.

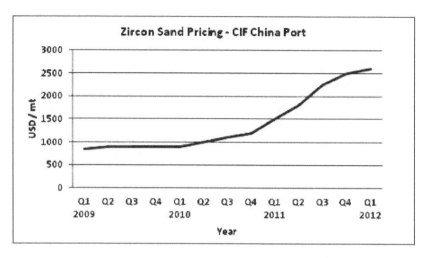

Figure 2. Zircon sand pricing (CIF China Port)

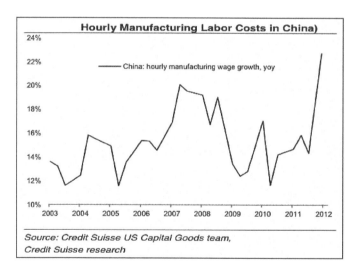

Figure 3. Labor cost in China over the past decade

While the insufficient improvement of quality and the almost inexistent development of services make it impossible dropping the price leverage, the manufacturing cost increase compresses the margin of contribution of Chinese fused cast (sold in the west) so dramatically to make these sales undesirable and almost financially unfeasible.

To understand what the problem of profitability for Chinese manufacturers is, you must take in consideration that the critical decision-maker, from the perspective of a western glassmaker, is the book cost of products and services delivered at their own premises.

ADDING COSTS TO THE PRODUCTS

While a western supplier "embeds" in the product price a number of pre- and after-sale services, and the transportation is in most cases within the same continent, in case of a low-cost sourcing, the product cost is only relevant to the materials, while all the connected services are, with variable levels of effectiveness, provided by a third party (agent, distributor, independent professional) or by internal organization of the glassmaker if this is big enough. Transportation is, generally, transcontinental. The cost of appended services, whoever will provide, and the differential in transportation, is adding an estimate five to ten percent to the product cost that, if a certain level of price leverage (e.g. 25%) must be maintained on a delivered basis, it has to be absorbed by the manufacturer and will further compress the margin of contribution. Figure 4 compares the cost addition by Western manufacturers vs. Chinese suppliers

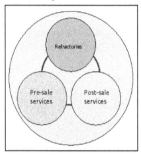 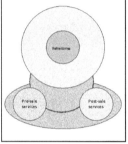

Western manufacturers supply   Chinese supply only products,
services embedded with products   third party provides services.

Figure 4. Cost addition by West manufacturers vs. Chines suppliers

Running some costing model, I have estimated that, in these days, the overall standard gross margin for Chinese manufacturers selling AZS fused cast in the west is rapidly going toward zero, being in most cases lower than the skimpy single digit margin of western players.

In general, we must say that western manufacturers, that went through downsizing, streamlining the organizations, cutting overhead costs and mainly reinforcing the servicing to Customers, have been quite effective in stabilizing the business structure and the customer portfolio. Chinese, who pointed primarily on the cost leverage, went through (and still are in) a financial deterioration leading to a very problematic business perspective.

In a business environment where a huge over capacity (mostly concentrated in China) will stay in the medium long term, where harder-than-China low-cost places are well on the way of emersion, what is the survival perspective of the more than twenty-five Chinese fused cast manufacturers?

SURVIVAL CONDITIONS AND A VALUE vs. COST PROPOSITION

In general terms, since the costing structure of Chinese products will inevitably evolve toward a model closer and closer to the western, so it will have to be for all other aspects. Chinese materials will have to get closer to world-level refractories for quality of materials and associated services, technical marketing and cooperative action teaming with glassmakers. It surely will be not a short run, but more likely a long learning curve, and possibly only a few players will survive the stress of this metamorphosis.

Within this scenario, other variables will determine the emergence of new products, not necessarily fused cast, that could reduce the global market of fused cast, to the point where even Chinese companies, traditionally not very concerned about break-even points, will have to come to terms with the huge over-capacity of their manufacturing configuration.

Glassmakers will play a key role, since the very existence of Chinese players is perceived as a value in keeping high the environment competitiveness.

While prices of low-cost refractories will rise on a purely survival stance, and, waiting for improvements to come, there will be a season when the ratio value/price will fall into a grey zone, where some glassmaker will be forced to reconsider the low-cost procurement policy.

Focusing on the value vs. cost proposition, we can distinguish different phases for past, present and future scenarios (as shown in Figure 5); since the advent of Chinese, low-cost, fused cast the ratio value/cost was made appealing by a really low cost, mostly supported by the cheap labor. Since then, and particularly in the last decade, the value did not greatly improve while the cost (sale price) came up relatively slow, maintaining a large and unchanged competitive delta over western pricing; in the recent period, though, prices have begun rising more, as a consequence of a large R.M. and labor inflation that cannot anymore be absorbed in term of reduced profitability.

In the short-term future, Chinese manufacturers will have to raise prices on a survival stance, and this will drop the ratio value/cost into a "grey zone" where some western glassmakers might put under scrutiny the low-cost procurement policy.

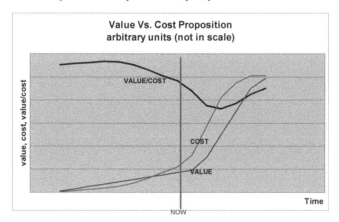

Figure 5. Value vs. cost proposition for past, present and future scenarios

The only way to recover the situation, will be rising the "value" term of the equation, that means improving products, delivering associated services, teaming with customers removing language and cultural barriers, removing the extra costs coming from third parties involvement. This implies a U-turn change versus the present purely commercial approach and will be feasible with a medium-long term learning curve before to get again into an appealing value per cost status: at that point, Chinese suppliers will be not anymore one hard "low-cost" option, and will compete on a more equal position against western rivals.

Glassmakers, at least the big ones, will inevitably consider the opportunity to help some selected Chinese fused cast manufacturer surviving and being able to provide, as soon as possible, the level of products and services quality that they need and expect.

Third parties, providing technical and technological support in an area between refractory makers and glassmakers, will accompany this transformation for a medium-long term, helping Chinese companies to survive through the evolution, and helping western Glassmakers to stay connected with a very imperfect set of providers, for the time necessary for them to improve and eventually fix most of the present deficiencies.

MONOLITHIC CROWN AND ITS BENEFITS, COLLOIDAL SILICA BONDED REFRACTORIES TECHNOLOGY

Ali Farhadi & Tom Fisher
Magneco/Metrel Inc.
Addison, IL & Pittsburg, OH USA

Alonso Gonzalez Rodriguez
Pavisa Group
Mxico City, Mexico

Mario Estrada
Vitro Group
Monterrey, Mexico

ABSTRACT

Magneco/Metrel, Inc. is the world leader of monolithic, no cement, colloidal silica bonded refractory technology. This refractory technology has been utilized for crown construction with many crowns in service and operations including melter, regenerator, oxy fuel melter and flu gas covers globally. This paper will provide a short review of the history of the monolithic crown and its construction. The presentation will focus on fundamental differences between monolithic crown and conventional brick/block crowns with regards to installation, maintenance, gas and alkali attacks, energy saving, overall performance and cost benefits. The principles of a monolithic crown can be applied to the brick/block crown by over coating the brick/block with a layer of monolithic refractory to seal joints.

In this paper we also focus on benefits and operational improvement that monolithic crown offers such as possibility of higher operation temperature, resistance to alkali gas attacks, elimination of the rat hole ,heat up and cool down consideration, heat loss reduction and energy saving. Case studies will detail the installation and performance of glass crowns utilizing this new unique technology.

INTRODUCTION

Magneco/Metrel, Inc. is the world's leading developer and manufacturer of no cement (CaO), colloidal silica bonded refractory monolithic technology, with headquarter in Addison, Illinois and facilities in 16 countries worldwide. The Pumpable family of nanotechnology products consists of pumpable and shotcrete formulations of colloidal silica bonded monolithic refractory products developed to meet and exceed the demands that different areas of the furnace requires.

Pavisa Group is a Mexican glass container company with 60 years of experience and leadership in the manufacturing and marketing of glass and crystal products for different industries including the wine, liquor, food, cosmetic, and pharmaceutical industries. All of the glass furnaces in Pavisa are 100% oxy-fuel system. Vitro Group is the leading glass manufacturer in Mexico and one of the world's major glass companies, backed by more than 100 years of experience in the industry. Specializes in Container and Flat glass, Machinery, Industrial equipment and chemicals. Vitro facilities produce, process, distribute and sell a wide range of glass products that offer solutions to multiple industries.

Many improvements have been made to reduce or minimize the corrosion and energy loss in glass furnace crowns. Demand of higher pull rate has led to higher furnace temperature and oxy-fuel firing which requires glass manufacturers to consider use of alternative refractory products and innovative crown construction techniques. Improvements in refractory material led to reduce the amount of calcium in silica crown and use of alumina base refractory in oxy-fuel furnaces. On the other hand understanding of corrosion mechanisms led to decrease or eliminate the joints between the bricks by using bigger size blocks and concept of monolithic crown construction.

Compared to alumina, silica is the preferred material for crown construction because of its lower cost, good strength and defect potential, lower density, which simplifies furnace construction and lower thermal conductivity. However corrosion of the silica brick of glass furnace crown is a problem in air-fuel fired burner and accelerated corrosion reported in oxy-fuel. The concentration of alkali gas (which in oxy-fuel fired system is two to four time higher than air-fuel fired) resulted unwanted corrosion in silica bricks [1, 2].

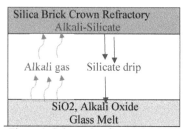

Figure 1. Schematic of corrosion mechanism process of silica brick refractory

Alkali Oxide $\longleftrightarrow$ Alkali,gas (OH, S)
(1350 °C, 2460 °F)
Alkali,gas + Silica $\longrightarrow$ Alkali-Silicate   (650-950 °C)   (1200 to 1740 °F)

Inside the glass melter alkali hydroxide (NaOH or KOH) or alkali sulfate ($NaSO_4$ or KSO_4) produced by reaction of combustion generated water vapor or sulfur gas with the alkaline oxides in the glass melt.

Corrosion mechanism can be explained by alkali gas penetration into the porous or joints with subsequent condensation of alkali gas to the external surface of the refractory. Temperature, gas velocity and partial pressure of alkali gas also affect the whole corrosion mechanism. Figure 1.

The chemical changes in this reaction zone cause liquid formation of alkali gas, then diffusion of the liquid into silica surface cause dissolution of silica to form a alkali-silicate. The amount of alkali-silicate increases with decreasing temperature of the refractory. Calculated phase diagram for alkali-silicate, sodium and potassium are show in Figure 2 [1].

Ratholes are formation of large voids from inside of the silica brick refractory, these voids are generally connected to hot face of the refractory via narrow channel. The initial formation of the channel may be caused by relatively slow corrosion most probably along the joints or at the defect of refractory

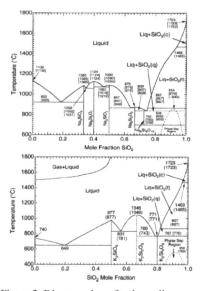

Figure 2. Diagram phase for the sodium-potassium-silica system

product. Therefore, liquid alkali-silicate either sucking into the porous, dripping down or flowing away.

It was suggested that insulation of the cold face of the refractory to minimize the temperature gradient. Since alkali-silicate increases with decreasing temperature of the refractory, so higher refractory temperature could prevent rathole formation (Figure 3) by promoting the formation of more stable silicate that are less prone to drip down or flow away from the surface.

For the same reason, in order to avoid the temperature gradient across the crown refractory thickness during furnace operation, it is recommended that all brick joints to be sealed upon early start of the furnace operation.

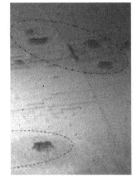

Alkali gases can travel to cooler portions of the refractory through the joints. So alkali gas penetration into joints with subsequent condensation and corrosion lowers the mechanical integrity of the crown [2, 3, 4].

During 1995-1998 Praxair and a consortium of glass companies funded a series of laboratory studies and thermodynamic modeling at TNO Institute of Applied Physics to explain the corrosion mechanisms of silica brick crown refractory (Figure 4) and to develop ways to reduce corrosion especially for oxy-fuel fired environment. The key finding of the studies are as follows [5]:

Figure 3. Rathole formation

a) most severe corrosion occurred at the joints of crown bricks,
b) the corrosion rate of silica bricks and joints are strongly related to the mass transfer of alkali gas into the bricks,
c) the calcium rich bonding phases (CaO-SiO_2) in silica bricks is preferentially attached by alkali gas, forming a glassy phase which penetrate through the silica grain boundaries and dissolves silica grains,
d) fused silica reacts less intensively with alkali gas,
e) fused silica bricks with a very low calcium content are preferred over the regular silica bricks for the crown construction,
f) pure alumina show very low corrosion rate for alkali gas

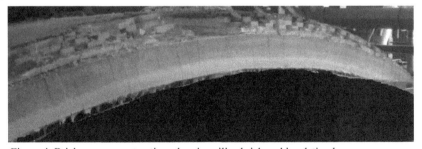

Figure 4. Brick crown construction, showing silica brick and insulation layers

Alumina-based refractory products provide alternatives to silica refractories. Un-like silica, liquid-phase alkali-aluminates are stable at very high temperatures and at extremely high alkali gas partial pressures. Thus, corrosion and rathole are less likely to form in alumina bricks crown at normal glass furnace environment. However alumina crowns are heavier, more expensive than silica and are also subject to varying degrees of corrosion in oxy-fuel fired environments [6].

BRICK JOINTS, ENERGY LOSS AND OVER COATING THE CROWN

It's already mentioned that corrosion reaction accelerates inside the joints due to temperature gradient through the joint (Figure 5). Furnace designer always try to minimize the joints and have very tight brick-laying to avoid such joints. Furnace operators continually try to close and cover the joints or any opening as soon as they form or appear in the brick crown.

Additionally joints and pores are the places where accumulated alkali gas can provide seeding spot for condensation reaction. Furthermore joints create venting inside the furnace, gas and heat can passes from inside to outside and vice versa. Thus addition to corrosion, joints cause energy loss from the system. Many improvements have been made to reduce or minimize the corrosion and energy loss. Figure 5 shows the example of the brick crown joint opening.

In the glass furnace, in general, about 33-40% of the energy goes toward melting the glass. Up to 30% of the energy consumed by a furnace can be lost through flue gas

Figure 5. Brick crown joints

exiting the stack, while another 30% can be lost through its refractory structure and the joints[7]. How much of the energy loss through the structure is actually cause by joint?

Any way crown refractory is subjected to corrosion, as refractory deterioration can lead to significant energy losses too. At the end of campaign life energy use can be up to 20% more than at the beginning of campaign life due to refractory lining loss [7]. Thereofore, a corrective and preventive measures overcoating is recommended for old and new brick/block crowns as shown in Figure 6. For new crown overcoating is to seal the bricks/blocks from any imperfection on the brick/block joints so early prevention from rathole formation. For the old brick/block crown overcoating give addition integrity and prolong the life of the crown, also to slow down the rathole formation and to decrease the energy loss.

Overcoated Crown   Not overcoated Crown

Figure 6. Overcoating

For a crown dimension of 10 meter (35 ft) length by 8 meter (26 ft) span, if assume 1/16 of inch opening between the joints between the bricks or blocks, and if consider the bricks dimensions of 3 inch by 6 inch or for block dimension of 19 inch by 35 inch. Calculating the opening space cause by joint for the bricks is 3% of the total surface and for the blocks is 0.5% for the total surface of the crown, how this open area affect the energy loss need to be evaluated.

Magneco/Metrel has proven data, Table 1. that shows monolithic construction of a glass furnace reduces the energy loss (these results are validated by monolithic application in other

Table 1. Different cases shows energy saving by monolithic application		Natural Gas Reduction
1	100% Pumpable, monolithic furnace lining	13%
2	Major tank and crown repair	13 to 19%
3	100% Pumpable, monolithic furnace lining	20%
4	Hot Crown construction (200Mm)	18%
5	Hot crown overcoating	12%

industry such as aluminum or iron and steel). The energy saving results are spread from 10% to 20% varying by the country, operational practice, thickness of the over coat, partial or major monolithic area [8].

COLLODIAL SILLICA BONDED TECHNOLOGY

Throughout the past 20 years, colloidal silica bonded monolithic refractories have been developed and utilized for their properties and ease of installation. The improved high temperature properties, abrasion resistance, thermal shock resistance, corrosion resistance, superior alkali resistance, and versatility in installation method make colloidal silica, no calciume bonded refractories an excellent alternative to traditional brick/block crown refractory.

Colloidal silica bonded refractory products shown in Figure 7 are based on sol-gel bonding technology and use a liquid binder system which contains no calcium (calcium-silicate-aluminate) and no chemically bonded water. All free water is available for easy removal at 100°C resulting in much less cure times or faster dry-out schedules.

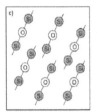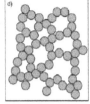

c) Siloxane bonds (-Si-O-Si-).   d) Three-dimensional network of SiO₂ particles

Figure 7. Colloidal Silica as a Nanostructured Binder

Colloidal silica bonded refractory products are also very "adhesive" and strongly bond to themselves or to other clean and non-degraded refractory substrates. The ultra-fine nature of the colloidal silica binder and the nature of the bond formation allows for excellent bonding to existing refractory linings. The nanometer-sized particles can easily penetrate the surface of existing linings. Siloxane bonds then form and then those bonds penetrate into the surface material of the lining. Figure 8. illustrates the bonding of a colloidal silica bonded castable to another refractory material and the excellent bonding that results. This bonding capability proves especially valuable because

Figure 8. Fine colloidal silica particle size promotes strong bonding properties

colloidal silica bonded refractory contain NO calcium, so they do not degrade after exposure to very high temperatures, as do all conventional castables.

MONOLITHIC CROWN TECHNOLOGY, PUMPABLE PRODUCT TECHNOLOGY

Magneco/Metrel's unique family of proprietary, monolithic, Metpump brand refractory products utilize colloidal silica (sol-gel) bonded, no-calcium, no added water technology. Magneco/Metrel Metpump pumpable system is delivered in 2 or 3 components: Metpump dry powder and liquid binder and/or accelerant.

The required refractory products, liquids, mixers, pumps are properly set-up and the required slurry delivery pipes and hoses then to be installed and their interior surfaces will be lubricated. Each batch of Metpump refractory is properly mixed and then each batch of mixed slurry are "Slump and Flow" tested for required characteristics. After approval each tested slurry batch is introduced into the refractory pump. The slurry then pumped directly to the product application points where it can be pneumatically shotcreted or pumped into place. Figure 9 shows pumping into place of the monolithic crown. Magneco/Metrel's standard operating procedures (SOPs) are in effect during entire preparation and application.

Magneco/Metrel reported Metpump product Met-Silcast have been performing extremely well in crown construction and overcoat repair as silica base refractory product since 2001. Met-Silcast has been used in more than 200 facilities around the world as monolithic crown, monolithic furnace structure, over coat crown or structure as hot or cold application. Met-Silcast is a cement free, high purity fused silica pumpable product. If applied as new monolithic crown Met-Silcast exhibits excellent strength, thermal shock resistance, and resistance to alkali attack. Met-Silcast if used as over coat substance, adheres to the silica brick crown and reinforces it, since Met-Silcast has very low thermal expansion rate, lamination is minimal so by closing the opening and joints reduces attack on the old silica crown, and therefore reduces the glass contamination danger.

Figure 9. Colloidal silica pumpable product

Table 2. Pumpable colloidal silicate bonded refractory products chemical composition. Monolithic upper form

Chemistry	Crown Pumpable Met-silcast	Conventional Silica Brick	Crown Pumpable Metpump C-190G	Crown Pumpable Metpump IPSXG
Al2O3, %	0.1	0.2	87.80%	65.7
SiO2 %	99.7	96.6	11.7	31.4
CaO %	0.1	2.9	0.1	0.1
Alkalis %	0.1	0.1	0.1	0.4
TiO2 %			0.1	1.5
Fe2O3 %			0.1	0.8
Bulk Density lb/ft2	115		180	160
HMOR at 1480°C	3.9 MPa		8.4 MPa	7.0 MPa

Magneco/Metrel also posted Metpump C-190 G has outstanding result in superstructure and crown construction as alumina base refractory [10]. Metpump C-190 G have reduced both cost and installation time, while providing superior performance compared to brick/block, thereby increasing service life and improving equipment availability. Metpump C-190 G is a high alumina calcium free pumpable product, it has excellent hot strength, thermal shock resistance and volume stability.

Forming a complete crown consist of construction of the lower form, similar to brick crown lower form. The lower form in monolithic crown need to be sealed and completely supported from lower area. Upper form could be metallic or wooden and has the same arch degree as lower form, it will be supported from upper structure. Similar to brick crown skew block will support the entire weight of the crown. The skew could be monolithic or blocks. Figure 10 shows the metallic upper form and Figure 11 shows the sketch of a monolithic crown construction.

Figure 10. Monolithic upper form

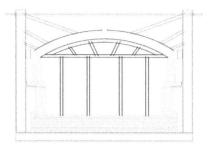

Figure 11. Monolithic crown construction

CASE STUDIES - MONOLITHIC VS BRICK CROWN

First case study is comparing Magneco/Metrel Met-Silcast monolithic crown vs. conventional silica brick crown construction (Table 3). Both cases are container glass furnace at 220 ton glass per day pull rate. Compare to conventional silica bricks, Met-Silcast contain fused silica grains, no calcium and excellent volume stability. Cost of Met-Silcast is about 26 percent more than silica bricks. So as for forming and installation cost 33% and 50% respectively. The main reason for this higher investment cost is higher premium refractory product technology and the fact that upper form is needed in case of monolithic application. Met-Silcast monolithic has no joint and also lower thermal conductivity so in general one or two layers less insulation brick is used compare to silica brick crown. Assuming one-layer insulation less, cost of insulation is 50% less in monolithic application. Now a day overcoating of the bricks is essential to sustain the long-term operation without the ratholes, however monolithic application does not need overcoating. Therefore in general investment cost on the first year is 8% more for monolithic compare to silica bricks.

We need to look at the saving at the end of the campaign life, in this case we assume 6 years for the campaign life. In reality during the campaign life could be many hot repair on the silica bricks due to brick work imperfections which it does not accrue in monolithic application. Energy saving in monolithic application is between 10 to 20 percent however we assume 10% energy saving in this exercise. One can determine total cost of

Table 3. Case study 1 – Container glass

Case study-1 Container Glass	Met-Silcast Monolithic Crown	Silica Brick Crown
Furnace capacity	220 tpd	220 tpd
Crown dimension	10 x 8 x .380 m	11 x 8 x .380 m
Raw material quality	Fused Grains	Calcined Grains
Density	1,840 Kg / M3	1,810 Kg / M3
Required material crown	61500 kg	7000 #
Volume Stability	Excellent	Poor
Met-Silcast and brick + mortar cost	26 % more	1$ bases
Form cost, both forms	33 % more	1$ bases
Installation cost, bricks, pump	50 % more	1$ bases
Insulating bricks	1$ bases	56 % More
Over coating, Seal Joint at 5-20 mm	1$ bases	100 % More
Investment Cost, First Year	8 % more	1$ bases
Hot Repair/maintenance, $14,800/year X 6 yrs	1$ bases	100 % More
Enegy cost, 22-107$/tgp, X 6 yrs	1$ bases	10 % More
Total cost of ownership in 6 yrs Material & Energy	1$ bases	11 % More

ownership during these 6 years, which in case of monolithic application is proven that about 11% less cost than brick crown. Which in this case, the cost savings will cover the size of the investment for complete furnace reline.

Second case study is comparing Magneco/Metrel Metpump C190G monolithic crown vs. conventional alumina block crown construction (Table 4). Both cases are container glass furnace at 20 ton glass per day pull rate oxy-fuel firing melter.

Cost of fused alumina bricks is about 40 percent more expensive than Metpump C190G. Forming cost is 25% more expensive for monolithic since upper form is needed in case of monolithic application. Installation cost in this specific case was 20% more for fused alumina blocks. Metpump C190G has less bulk density and has no joint, we used one layer insulation less, so cost of insulation was 67% more for alumina blocks. Originally alumina blocks were over coated which in case of monolithic it does not apply so positive cost saving for monolithic application. Consequently investment cost on the first year is 33% more for alumina blocks compare to monolithic application.

Looking at the campaign life, in this case we assume 6 years for the campaign life. We assume few hot repair on the alumina block due to brick work imperfections which it does not accrue in monolithic application. Energy saving in monolithic application is between 10 to 20 percent however we assume 10% energy saving in this exercise. Calculating the total cost of ownership during 6 years monolithic application expected to be 10% less cost than brick crown.

Table 4. Case study 2 – Oxyfuel furnace

Case study-2 OxyFuel Furnace	Metpump C190G Monolithic Crown	Alumina Block Crown
Furnace capacity	20 tpd	20 tpd
Crown dimension	4 x 2 x .30 m	4 x 2 x .30 m
Raw material quality	Metpump C190G	High alumina
Density	2930 kg/M3	3400 kg/M3
Metpump C190G or (brick + mortar) cost	1$ bases	40 % more
Form cost, both forms	25 % more	1$ bases
Installation cost, bricks, pump	1$ bases	20 % more
Insulating bricks	1$ bases	67 % More
Over coating, Seal Joint at 5-20 mm	1$ bases	100 % More
investment cost	1$ bases	33 % More
Hot Repair/maintenance, $14,800/year X 6 yrs	1$ bases	100 % More
Enegy cost, 22-107$/tgp, X 6 yrs	1$ bases	10 % More
Total cost of ownership in 6 yrs Material & Energy	1$ bases	10 % More

CONCLUSIONS

We have come a long way in this attempt to improve the refractory product technology and application innovative methods. Eliminating calcium from ingredients of refractory product and overcoat or seal brick joints and eventually monolithic crown refractory technology. Overcoating of an old crown decreases gas venting, lower alkali gas attack and lessens the energy loss through the system, and it gives strength and integrity to the whole structure. Met-Silcast has proven result when overcoating silica brick crown and Metpump C190G has the same success story for alumina block crown. Monolithic crown construction is a sustainable solution for glass furnace crown construction. Monolithic properties, no-joint concept eliminates the gas attack and rathole seeding from the start of the campaign life. Energy saving is proven in this concept. Additionally monolithic colloidal silica, no-calcium bonded technology, such as Met-Silcast or Metpump C190G products, provide exceptional physical and chemical properties, which justify viable solution for future constructions.

ACKNOWLEDGMENT

The authors wish to thank Santiago Suarez for his contributions to this paper.

REFERENCES

1. M.D. Allendorf and K. E. Spearb, Thermodynamic Analysis of Silica Refractory Corrosion in Glass-Melting Furnaces, J. Electrochemical Soc., 2001.
2. A. Balandis and D. Nizeviciene, Silica Crown Refractory Corrosion in Glass Melting Furnaces, Kaunas University Of Technology, Kaunas, Lithuania Science of Sintering, 2010.
3. M. Velez, J. Smith and R. E. Moore, Refractory Degradation in Glass tank Melters. A Survey of Testing Methods, University of Missouri-Rolla, Department of Ceramic Engineering, Rolla, Missouri 65409-0330, USA
4. M.D. Allendorf , K.E. Spear, et. al., Analytical Models for High-temperature Corrosion of Silica Refractories in Glass-melting Furnaces, J. Electrochemical Soc., 2001.
5. H. Kobayashi, Advances in Oxy-Fuel Fired Glass Melting Technology, XX international Congress on Glass (ICG), Kyoto, Japan, Sept 26-Oct. 1, 2004.
6. K.E. Speara and M.D. Allendorf, Thermodynamic Analysis of Alumina Refractory Corrosion by Sodium or Potassium Hydroxide in Glass Melting Furnaces, J. Electrochemical Soc., 2002.

HIGH EMISSIVITY COATINGS IN GLASS FURNACES

Tom Kleeb
North American Refractories Company
Pittsburgh, PA

Bill Fausey
Owens Corning
Granville, OH

ABSTRACT
High emissivity coatings have been used in glass melters for over five years. To date, furnaces of over two dozen glassmakers have used these coatings. In most cases, they have resulted in significant fuel savings, more even heating, more rapid furnace commissioning, and a lower carbon footprint. In a few cases, the performance of these coatings has not been satisfactory under certain furnace and operating conditions. An Owens Corning study of insulation furnaces using high emissivity coatings has shown consistent performance between furnaces and fuel saving benefits in excess of five years.

BACKGROUND
Although high emissivity coatings have been used in petrochemical and other industrial applications for decades, the coatings available were not suitable for use in the high temperatures in glass furnaces. High emissivity technology developed by NASA to protect the space shuttle during re-entry was licensed to Emisshield, Inc., which developed a family of high emissivity coatings having service temperatures of 1700°C and above[*].

Test panel trials of these coatings were installed and monitored in 2006 and 2007. Finally, the first commercial installation was made in Owens Corning's K5 furnace in Kansas City in 2008. Owens Corning first disclosed the use of a high emissivity coating in the insulation furnace at the GMIC energy workshop of 2009. This was followed by a paper as part of the 71st Conference on Glass Problems[1] summarizing the energy savings in the 150 metric ton per day oxy-fuel furnace. Both of these presentations clearly demonstrated the energy saving opportunity of the coating.

HIGH EMISSIVITY COATINGS
High emissivity coatings are different from reflectors in that they always absorb convective and radiative heat on their surfaces. If there is not a colder body to radiate to, the absorbed heat will be conducted through the coating to the refractory to which it's applied, increasing the temperature of the refractory. If a colder body is available, such as batch or cooler molten glass, the coating will radiate the heat to that body as a complete black body spectrum. The effect of using a high emissivity coating is to increase the radiant heating component in a glass furnace at the expense of convective heating. The combustion gases in the furnace atmosphere are capable of absorbing specific wavelengths of the spectrum (Figure 1), so the remaining wavelengths are re-radiated as heat that passes through the gases and is absorbed by the batch or glass. The effect of this is that the glass surface may be hotter than the furnace gases.

---

[*] EMISSHIELD® is a registered trademark for coatings manufactured by Emisshield, Incorporated, and is covered by U.S. Patent 6,921,431

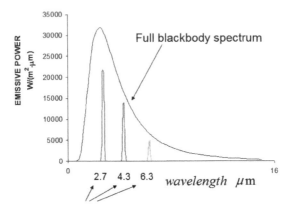

Figure 1. The full black body spectrum emitted by the high emissivity coating along with the major wavelengths absorbed by carbon dioxide and water vapor in the combustion gases.

The energy re-radiated from a high emissivity coating and absorbed by batch and glass can be described by the following equation:

$$Q = E_w \cdot \sigma \cdot (T_C{}^4 - T_L{}^4)$$

Where: Q = re-radiated energy absorbed by the furnace load
E_w = emissivity of the coating
σ = Stefan-Boltzmann constant
T_C = coating temperature
T_L = load (glass) temperature

It is obvious that energy will be radiated from the coating as long as there is a temperature difference between the hotter crown and the surface of the furnace load, the batch and the glass. In most fiberglass and soda-lime furnaces, whether they use high emissivity coatings or not, the crown and superstructure refractories radiate heat to the glass. Most alumina and silica refractories have emissivities of about 0.5; refractories containing chromia or zirconia have somewhat higher emissivities of about 0.65. By applying a high emissivity coating to these refractories, the emissivity of the radiating surface is increased to at least 0.9. Mathematically, the equation describing energy radiation looks like this, where the increased emissivity causes the amount of radiated energy to increase:

$$Q = E_w \cdot \sigma \cdot (T_C{}^4 - T_L{}^4)$$

The increase in radiant energy from the coated refractories is not necessarily desirable. A hotter surface temperature may result in hotter glass at the throat and a change in sub-surface convection patterns. It is usually preferred to restrict the effects of using a high emissivity coating to the furnace above the glass surface, keeping conditions within the glass bath and in the process downstream, unchanged. In order to do this, the burners must be turned down until Q returns to its pre-coating value. With Q at its original value and E_w increased due to the application of the coating, the only way the equation can be balanced is by reducing the temperature of the radiating surfaces, T_C:

$$Q = E_w \cdot \sigma \cdot (T_C{}^4 - T_L{}^4)$$

Since the radiative heating of the glass surface has been maintained at the pre-coating level while burning less gas, fuel savings are attained.

COMMERCIAL HIGH EMISSIVITY COATING UPDATE

At the time the authors last presented a paper on high emissivity coatings three years ago, fifteen furnaces melting C-Glass, E-Glass, soda-lime glass, and borosilicate lighting glass had been coated. These furnaces were operated by six North American glass makers. Since then, twenty-six glass companies operating sixty-eight furnaces in the Americas, Europe, and Asia are using high emissivity coatings (Figures 2 and 3).

Figure 2. High emissivity coating being applied to zircon refractory in an E-Glass furnace

Figure 3. Completed C-Glass furnace crown and superstructure installation

These furnaces have been about evenly divided between gas/air and oxy-fuel firing systems. In most C-Glass furnaces, the chrome crowns and superstructures have been coated. In E-Glass furnaces, soda-lime container and tableware furnaces, and furnaces producing borosilicate, lighting, and sodium silicate glasses, silica, mullite, alumina, zircon, and bonded AZS refractories have been coated. One notable refractory that has not been coated is fused AZS. The glass that exudes from fused AZS in service prevents high emissivity coatings from adhering to these refractories. For this reason, soda-lime furnaces constructed with fused AZS refractories in the superstructure have only their crowns coated. There have been no commercial installations of high emissivity coatings on fused beta-alumina refractories, but some of these coatings appear to adhere well under laboratory conditions.

To date, North American Refractories has installed five high emissivity coatings from two manufacturers. The coating selection is largely dependent upon the composition of the substrate being coated, as binder systems have been optimized for different substrates. Other coating selection factors include the condition of the refractory to be coated and the physical location of the furnace. Some coatings are not available worldwide.

BENEFITS OF USING HIGH EMISSIVITY COATINGS

After a high emissivity coating is installed, the first effect in furnace operation will become apparent during the heat-up. The coating will absorb and re-radiate to the batch and cullet some of the energy that normally heats the refractories. Proof of this can be seen in Figure 4, which shows a test panel of high emissivity coating that was sprayed on to a silica crown of a container furnace. The test panel, which is near the regenerator ports, was measured by optical pyrometer to be about 35°C cooler than the uncoated crown. Since the coating surface is cooler, the refractory behind it is cooler and more heat is directed to the batch. This results in faster heat-up, if desired, or a standard heat-up using less fuel. One high emissivity coating user claims that the fuel savings on heat-up exceeds the cost of the coating.

Figure 4. Cooler crown test panel next to regenerator ports showing lower luminosity compared to the uncoated portion of the crown

Most users of high emissivity coatings expect to offset the relatively high cost of the coating with fuel savings. Unlike the use of insulation which prevents heat loss through a multicomponent refractory lining, high emissivity coatings reduce heat loss up the stack by directing heat to the glass being melted. There are many variables affecting the efficiency of this process, some of which will be discussed in the next section of this paper. The sheer number of variables affecting the performance of high emissivity coatings make calculating the fuel savings of a glass furnace difficult. However, after over five years of using high emissivity coatings in glass melters, some trends have emerged. When the refractory design of a furnace permits the crown and superstructure to be coated, most furnace operators report a 6% to 8% fuel savings. When only the crown is coated, as is the case for furnaces with fused AZS superstructures, 3% to 4% fuel savings is typical. Whatever the fuel savings experienced, CO_2, and in the case of gas/air furnaces, NO_X emissions are reduced proportionately. Clearly, the use of high emissivity coatings can help glassmakers achieve environmental impact goals.

FACTORS ADVERSELY AFFECTING THE EVALUATION OR PERFORMANCE OF HIGH EMISSIVITY COATINGS

While most users who have experienced fuel savings in the ranges listed in the previous section have seen a payback on the cost of the coating in less than one year, other users have experienced disappointing savings or none at all. There are a number of reasons for this, some of them operational and some of them process related, that could lead to apparently poor coating performance. Some of these are listed below:

- For first-time high emissivity coating users, it is generally not a good idea to try to evaluate them in new furnaces. Without baseline data collected on an uncoated furnace melting the same glass at the same cullet level and at the same pull rate, there is no way of quantifying the benefits of the coating.
- Similarly, a furnace with an extensive performance history that underwent a major rebuild will have performance improvements as a result of that work. If a high emissivity coating is applied before heat-up, separating the performance improvement due to the coating from the improvement due to the repairs may prove difficult. Often, glass companies have historic expectations of performance improvement due to the rebuild components and any incremental improvement beyond that is attributed to the coating. The evaluation of a high emissivity coating under these conditions is problematic.
- One of the major causes of apparent poor performance of high emissivity coatings is an operator's unwillingness to deviate from the standard operating protocol. The proper use of these coatings will leave conditions below the glass surface and downstream in the process unchanged. Above the glass surface, everything will change. The furnace atmosphere will be cooler, the hot face of coated refractories will be cooler, burner settings will be turned down, and control thermocouples drilling into refractories behind the coating or drilled through the crown will detect lower temperatures. If the furnace is controlled by a thermocouple that is measuring the temperature of the atmosphere, actual fuel usage may increase.
- In the past three years, an E-Glass producer who normally runs their furnaces with an unusually thick layer of foam on the surface experienced no fuel savings when the coating was applied to two of their furnaces. It is thought that the foam absorbed the energy radiated from the coating, but instead of conducting it to the glass, re-radiated it back to the furnace atmosphere. Before installing a high emissivity coating, it is imperative that the design and operation of the furnace is fully understood.

The economic use of high emissivity coatings requires an understanding of emissivity and the operational parameters of furnace candidates to use the coating. Owens Corning was the first glassmaker to use high emissivity coatings. They have good baseline data on the furnaces before they were coated and have been monitoring fuel usage regularly.

HIGH EMISSIVITY COATING ON OWENS CORNING INSULATION FURNACES
After the two presentations referenced in the BACKGROUND section of this paper were made, two important questions remained to be answered. First, how long would the coating survive in the furnace atmosphere and second, if the coating remained in place would the energy savings continue over several years. Since these early disclosures, Owens Corning has coated four additional oxy-fuel furnaces and two air-gas/electric boost furnaces. The coating has shown a consistent energy saving in all applications and several years of operation are beginning to offer insight into the two questions.
The energy savings demonstrated in five oxy-fuel furnaces with high emissivity coating in place is shown in Table 1. These furnaces have a thru-put ranging from 80 metric ton/day to 150 metric ton/day. Clearly the energy savings is repeatable and consistent. These data indicate the energy savings from the coating installed on K-5 furnace is decreased in year five of operation. There were no changes to the batch composition or furnace operation that would explain this reduction in energy saving. At this time it is not clear if this represents a new plateau that will hold for several years or if this is the beginning of a reduction to zero energy savings. Importantly, several furnaces will cross this five year time period over the next several years at Owens Corning. This will allow for a larger data set to be obtained for the evaluation of effective life of the coating and energy savings.

	2008	2009	2010	2011	2012
K-5	7-8%	7-8%	7-8%	7-8%	4-5%
DM-1		7-8%	7-8%	7-8%	7-8%
V-1			7-8%	7-8%	7-8%
DM-2				7-8%	7-8%
MC-1					7-8%

Table 1. Energy savings by year in five Owens Corning oxy-fuel insulation furnaces

With K-5 furnace showing the decrease in energy savings in year five, one might ask if this is simply the energy efficiency deterioration of the furnace due to age and as the furnace becomes less efficient then the measured savings go down. History of energy monitoring of the oxy-fuel furnaces at Owens Corning does not show a decrease in energy efficiency at the five year mark. At this time it is believed the energy efficiency loss is due to deterioration of the coating but this will be evaluated over the next several years as K-5 and the other oxy-fuel furnaces continue their furnace campaigns.
The application of a high emissivity coating to the entire superstructure of oxy-fuel and gas fired furnaces continues to be standard practice at Owens Corning. Knowing that significant energy saving is still being realized after five years of operation makes the use of the coating a sound melting and business decision.

FUTURE DIRECTIONS

High emissivity coatings have now been installed in furnaces in the Americas, Europe, China, and Southeast Asia. Initiatives to offer these products to glass producers in un-served markets, such as Japan, India, Australia, and the Middle East have begun. Binder systems of the existing product line are being refined and a trial of a new, higher-performance coating is underway. New installation techniques, including a method for installing these coatings in operating furnaces, are being studied. Trials in float glass furnaces have indicated that high emissivity coatings perform differently in these furnaces. These trials are too early to draw any conclusions regarding the use and effectiveness of high emissivity coatings in these furnaces.

REFERENCES

1. Kleeb, Tom & Fausey, Bill, "Fuel Savings with High Emissivity Coatings," 71st Conference on Glass Problems, October 19-20, 2010, Columbus, Ohio, Pages 125-136.

ACKNOWLEDGEMENT

The authors wish to thank Don Shamp and Fuse Tech for their contribution of Figure 4 to this paper.

NEW RECYCLING SOLUTION FOR REFRACTORIES FROM INSULATION GLASS FURNACES

O. Citti[*]

C. Linnot[**]

T. Champion[***]

P. Lenfant[***]

[*]Saint-Gobain NRDC, Northboro, MA01532

[**]Savoie Refractaires, Venissieux, France

[***]Saint-Gobain CREE, Cavaillon, France

ABSTRACT

SEFPRO and its subsidiary VALOREF have offered solutions for the management of refractory wastes issued from total or partial repairs of glass furnaces for more than 30 years. To improve the sustainability of our chrome bearing refractory for the insulation glass industry, a new solution has been developed for the complete recycling of refractories containing Cr_2O_3 from the insulation glass industry and has started to be deployed worldwide. The recycling of Cr bearing refractory is far more challenging than for other refractories due to the potential formation of hexavalent chromium during the use of these refractories in the alkali environment of an insulation glass furnace operation. Contrary to current selective recycling process offered on the market, SEFPRO has decided to invest in an original recycling process which allows 100% of Cr waste to be treated, while complying with all current international EHS standards. After dismantlement of the furnace, used Cr refractories are decontaminated, processed and then recycled within a new range of Cr containing refractory materials. Due to our new and patented process and refractory formulations, the new materials are designed to offer a high level of performance perfectly suited to the demands of the insulation glass industry, and in particular, high corrosion resistance, low Cr sublimation, adapted mechanical properties and high electrical resistivity.

INTRODUCTION

Among refractory oxides, Cr_2O_3 has the highest corrosion resistance to most glasses. In most applications, Cr_2O_3 refractories have more than twice the life compared to conventional AZS refractories, which makes them the best suited materials to resist high wear conditions. Our laboratory testing of the corrosion resistance in insulation glass, shows that a 30% Cr_2O_3 rich AZS (Alumina Zirconia Silica) material is 90% more corrosion resistant than a regular 32% ZrO_2 AZS material (figure 1). As the Cr_2O_3 content increases, the corrosion resistance increases dramatically. As a consequence, AZS-Cr materials are used in Insulation glass furnaces for the whole tank, but also in high wear areas of sodalime glass furnace (ZIRCHROM 30, ZIRCHROM 50, ZIRCHROM 60 and ZIRCHROM 85). Finally, materials containing more than 90% Cr_2O_3 (like SEFPRO C1221, C1215Z and C1215 materials) are used for reinforcement fiber-glass, which are high temperature and aggressive glasses with high glass quality requirements.

As a matter of fact, the use of chromium oxide refractories has rapidly become a key design solution for most insulation glass furnaces and reinforcement fiber-glass furnaces with no industrially viable alternative at the moment.

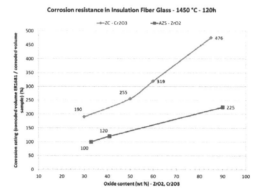

Figure 1. Comparison of the corrosion resistance of AZS-Cr materials in insulation glass at 1450 °C as a function of % Cr_2O_3 for Cr-AZS material and % ZrO_2 content for AZS materials.

However, in contact with alkali containing glasses, such as insulation fiber-glasses or sodalime glass, Cr_2O_3 containing refractories are likely to form soluble hexavalent chromium compounds during use. Substantial concentrations of sodium, potassium and/or calcium chromates have been measured in used Cr_2O_3 bearing products from furnaces having melt insulation glass or soda lime glass.

Cr_2O_3, the trivalent chromium form, which is used to make refractories, is considered not classifiable as to its carcinogenicity to humans by the IARC (International Agency for Research on Cancer) and other official organisms.

On the contrary, hexavalent chromium species (such as sodium chromate) are considered to be toxic and are classified as Carcinogenic to humans – Group 1 by the IARC as soon as 1990. Hexavalent chromium is classified in Europe as CMR (Carcinogenic : cat 1, Mutagenic : cat 2 and Reprotoxic : cat 3) and in the US it is also considered as Carcinogenic – Group A1 by the ACGIH (American Conference of Industrial Hygienists) since 1994. The toxicity for the environment is also known to be phytotoxic and causes environment problems. For those reasons, numerous regulations exist worldwide to control and limit the exposure to hexavalent chromium in the workplace, the emissions of Cr dust from industrial facilities, as well as the management of wastes containing hex chrome.

While recycling solutions exist for most non Cr-bearing solutions, as well as for spent isopressed Cr_2O_3 refractories from non-alkali glasses such as E glass, there's no full recycling program available for spent Cr bearing refractories from the insulation glass-wool industry. Current practice is to completely dispose of this dangerous waste and only recycle or reuse the non-dangerous waste. This situation leads to the disposal of a significant portion of the refractory wastes into approved landfills, often after treatment. It is also expensive and only partially relieves the glass makers of their legal responsibility towards the dangerous wastes that their process has generated.

In an effort to offer the glass-wool industry a real and global solution to the management of Cr containing wastes, as well as helping our customers alleviate their legal responsibility regarding the management of the industrial Cr wastes, Saint-Gobain SEFPRO and its subsidiaries Savoie Refractaires and VALOREF invested significant resources and funds to develop a 100% chrome refractory recycling program. Due to the multiple technical and industrial challenges that had to be addressed before a viable industrial solution could be put together, this R&D effort was pursued for more than 5 years. We are now exited to announce the upcoming availability of a 100% recycling solution for spent chrome refractories from the glass-wool industry in most countries around the world. Our solution includes the dismantling of the furnaces by Saint-Gobain VALOREF in Europe or our partners, the transportation of the wastes, their decontamination, and further processing for reintroduction in a newly developed and optimized refractory offer for the glass industry: WOOL 30 and WOOL50.

1. Brief description of the formation of hexavalent chrome from Cr_2O_3 rich refractories in glass furnaces environments

The formation of hexavalent chromium oxide in an alkali rich environment has been extensively studied in the past[1] and more recently in our R&D facility (Saint-Gobain CREE). One proposed mechanism for chromate formation from the chrome refractory in a glass furnace is often described by the following reaction:

$\frac{1}{2}$ Cr_2O_3(s, refractory) + Na_2O (g or l, glass) +3/4 O_2 (g) → Na_2CrO_4 (s,l,g)

While the formation of hexavalent chrome from refractories is not the focus of the work presented here, it is worth noting that the formation of hexavalent chrome is characterized by the following features:

- It is favored by oxygen rich environments, such as the glass line of the glass furnace or superstructure applications;

- It needs the presence of alkali or earth-alkalis to form and remain after cooling down;

- The formation is activated at temperatures above 600°C and increases with temperature;

- The physical form of the hexavalent chromium compound formed varies with its chemistry and the temperature: for example, sodium chromate is solid at low temperatures, melts at 796°C, and while it mostly remains liquid until 1200°C, it also vaporizes as the temperature increases.

The different physical forms of hexavalent compounds probably contribute to the increase in the hexavalent chrome contamination of the chrome refractory bricks with multiple phase migration mechanisms.

On the other hand, hexavalent chromium compounds such as sodium chrome are highly volatile at high temperature leading to the contamination of the fumes and the subsequent need for dust emissions abatement technologies.

Coming back to the contamination of the Cr_2O_3 rich refractories, we have surveyed a large number of used refractories of different origins in the furnace as well as different brands, and found significant but variable hexavalent chromium content in the spent refractories.

2. Disposal of spent chromium oxide refractories from the insulation fiber glass industry

In the US, RCRA 40 CFR part 261.3 (1976) requires that TLCP (Toxicity Characteristic Leachate Procedure, method EPA1311) must be performed to evaluate the toxicity (total Cr content) of any granular waste (up to 9 mm) containing Cr_2O_3. If Cr totals (TLCP) < 5 mg/l and no other pollutant is above limit, then disposal is possible as a regular industrial waste. On the contrary, if Cr totals (TCLP) > 5 mg/l then the wastes must be declared as hazardous and must meet the land ban standards (TLCP < 0.6 mg/l) before it can be landfilled. Actually, the EPA currently requests that, in addition to total Cr analysis,

---

[1] Formation of chromates from the reaction of alkali chlorides with Cr2O3 and oxygen – C. Hirayama, C.Y. Lin – National Bureau of Standards Special Publication 561, Proceedings of the 10th Materials Research Symposium on characterization of High Temperature vapors and Gases held at NBS, Gaithersburg, Maryland, September 18-22, 1978. Issued October 1979.

laboratories also determine hexavalent chromium content in the extract by the colorimetric method using diphenylcarbizide (DPC) as an indicator (EPA 7196A).

In Europe, the commission decision of May 2000 (2000/532/EC) pursuant to Article 1 of the Council Directive 75/442/EEC on waste, states that spent lining of glass refractories (category # 10-11-08) are considered to be dangerous waste when (not limited to) they contain more than 0.1 wt % of hexavalent chrome. Norm 15192,2006 describes the determination of chromium (VI) in solid material by alkaline digestion and ion chromatography with spectrophotometric detection. Compared to the TLCP for wastes, this method differs by a -250 micron particle size requirement for the powder and a temperature for extraction at 92.5 °C for 60 min. To be on the safe side, we actually developed a much more severe analytical method in our R&D center in CREE, to try to better estimate the total hexavalent chromium content of our refractories and have applied this method to spent refractories. While our production shows typical hexavalent chromium contents much below 100 ppm, the spent refractories varied between 50 ppm and 1 wt %, which confirms the dangerous characteristic of these wastes.

Norm En 12457, 2002 describes the compliance test for leaching of granular waste (up to 4 mm) materials (two Liquid/Solid ratios are used) in view of disposal in authorized landfills in Europe. The classification of wastes based on Cr leaching solely can be summarized as followed (see Table 1).

Table 1. Cr content limits for landfilling in Europe using norm 12457, 2002 (AFNOR : NF X 30-402-2).

Cr total (mg extracted / kg of waste)	L/S = 2 l/kg	L/S = 10 l/kg	Co (percolation)
Dangerous wastes not authorized for landfill = treatment needed	$Cr_{total} \geq 25$ mg/kg	$Cr_{total} \geq 70$ mg/kg	$Cr_{total} \geq 15$ mg/kg
Dangerous wastes = Class 1 landfill only	$Cr_{total} < 25$ mg/kg	$Cr_{total} < 70$ mg/kg	$Cr_{total} < 15$ mg/kg
Dangerous wastes = Class 2 landfill only	$Cr_{total} < 4$ mg/kg	$Cr_{total} < 10$ mg/kg	$Cr_{total} < 2.5$ mg/kg
Inert wastes = no restrictions	$Cr_{total} < 0.2$ mg/kg	$Cr_{total} < 0.5$ mg/kg	$Cr_{total} < 0.1$ mg/kg

Based on our survey, we can report that TCLP values of hexavalent chromium in spent chromium oxide refractories can often vary between several mg/l and several hundred mg/l depending on the position in the furnace and other considerations. Figure 2 shows an example of TLCP (EPA1311) values for Cr in chromium oxide refractories depending on their position in the kiln. It is worth noticing that the areas with maximum hexavalent chrome enrichment are the glass line of tank walls and superstructure materials, which is quite coherent with the previous discussion. On the other hand, we observed a higher level of hexavalent chrome formation for high chromium oxide content materials (for instance for the throat), than for low chromium content materials (pavers). Nevertheless, it is quite clear that spent chromium bearing refractories from insulation fiberglass furnaces must be considered as hazardous wastes. Their management is therefore submitted to many environment regulations, and incurs significant disposal costs.

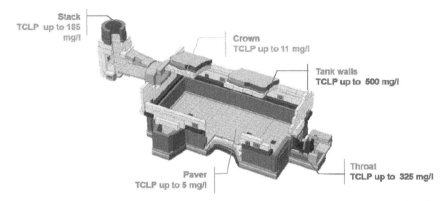

Figure 2. Simplified map of the hexavalent chromium oxide content of spent refractories after dismantling of an insulation glass furnace. The hex Cr contents are measured here using the TCLP (Toxicity Characteristic Leaching Procedure) method used for assessing the toxicity of the waste before landfilling.

3. SEFPRO Recycling solution of Cr refractories

The presence of hexavalent chromium makes the recycling of spent chromium oxide refractories difficult, therefore the development of our recycling solution has required several years of research and development. SEPR, being a leader in fusion cast technology, first investigated the use of a fusion cast process for the recycling of the spent refractories. However, we quickly recognized that such a process would require a hex Cr decontamination step, which was found to be expensive and challenging in terms of occupational safety. Consequently, the use of a fusion process was abandoned and a more traditional ceramic process, although very innovative, was designed and is now being industrialized in order to offer a sustainable 100% recycling process.

In our proprietary and patented process, the spent refractories are sorted and crushed on the glass plant site and then shipped to our VALOREF site in France. VALOREF/ALFAREF has a long experience of dismantling of insulation glass-wool furnaces and has developed specific EHS strategies adapted to hexavalent chromium oxide contaminated refractories.

Once received, the wastes are decontaminated and processed (proprietary method) to produce hexavalent chrome free raw materials. These reclaimed raw materials are then used in combination with fresh raw materials to produce high quality sintered grains, which are essential to the manufacturing of our Cr bearing refractories. Starting in 2014, SEFPRO will propose this recycling solution for Cr refractory wastes in Europe, as well as in the USA.

Due to the fact that dismantling the furnace is a key step in enabling full recycling, our offer includes the dismantling of the furnace by VALOREF in Europe and by SME in the USA. It is essential to make sure that proper glass removal is performed before the dismantling is started. Note that our threshold for glass contamination is as follows: less than 2 cm of glass on top of the pavers, and less than 1 cm of glass for the melter, throat and fore-hearth walls. It is also necessary to thoroughly sort the used refractory blocks during the dismantling, in order to identify and remove any refractory having less than 15% Cr_2O_3 (zircon, AZS, alumina, mullite, etc), but also to avoid any contamination from other types of wastes (insulation bricks, metal scraps, etc.).

Once the dismantling is completed, VALOREF in Europe or SME in the US take care of the transportation to our recycling facility (VALOREF) in France. In Europe[2], the transportation of the dangerous waste can be done under notification according to the Basel convention (1989). Even though the USA did not ratify the Basel convention, the transport of these dangerous wastes from the USA to Europe is authorized within the OCDE, provided the wastes can be fully reclaimed. The transportation must also be done after notification to the EPA and to the French authorities. Once the spent refractories are recycled, a certification of recycling can be issued to the glass maker.

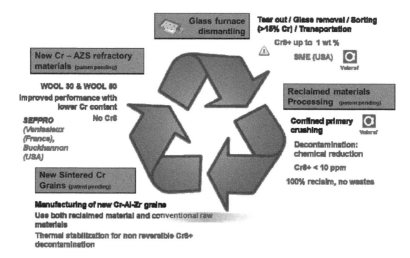

Figure 3. SEFPRO recycling program for Cr bearing spent refractories.

After the spent refractories are received by our Bollène facility, processing can start. The first process step consists of crushing and further milling in a confined facility built for this purpose under strict official authorization. Compared to other solutions available on the market, our proprietary[3] method consists of two steps. The first step allows us to chemically reduce the hexavalent chromium compounds to traces (compared to the original hexavalent chrome content). Even though this decontamination is still reversible under high temperature and oxidizing conditions, it enables further processing. The second step fully and definitely reverts the hexavalent chromium to the safe trivalent form through the addition of a proprietary combination of adjuvants and fresh raw materials, followed by a forming step and finally a high temperature firing to sinter high-density Cr_2O_3-Al_2O_3-ZrO_2 grains. Upon completion of this process, the grains produced do not contain any significant hexavalent chromium content. This second step is carefully managed to maintain the properties of grains no matter what the nature and chemistry of the spent refractories were at the beginning. Finally, the sintered grains can be used as grog material for the production of a new range of Cr bearing refractories for the glass industry.

To summarize, our recycling process offers:

• No disposal = Recycles 100% hazardous chromium refractory wastes if properly dismantled

---

[2] Except Russia who has not ratified the Basel convention and is not a member of the OCDE.
[3] Cf. patent application WO2012110952: METHOD FOR MANUFACTURING REFRACTORY GRAINS CONTAINING CHROMIUM(III) OXIDE,.

- Eliminates related landfill and wipe out landfill costs
- Reduced overall process wastes
- Saves time by offering complete coordination and paperwork
- Limited to refractories containing > 15% Cr_2O_3
- Accepts other suppliers materials (SERV, RK, etc.)
- Worldwide recycling (OCDE, Basel) approved by EPA and INERIS
- Available in the US and Europe, starting in 2014 and, most importantly,
- Fully respects our environment and offers a sustainable solution for the use of chromium oxide refractories.

4. Development of a new range of refractories for the insulation glass fiber industry

The Zirchrom materials range was developed more than 30 years ago by SEPR and was rapidly adopted by the industry due the significant capital cost reduction it allowed. The commercial development of this range of materials was strongly supported by manufacturing capabilities at our Savoie facility in Vénissieux (France), but also at our US facility in Buckhannon (WV). The Zirchrom range was AZS – Cr materials and contained high proportions of fused cast grains (AZS and AZS-28%Cr). In addition to their high corrosion resistance, these products were also characterized by a high level of thermal shock resistance, due to the introduction of monoclinic zirconia[4] in the batch material (cf. ZIRCHROM 85).

However, our current range of materials doesn't allow the use of reclaimed material without a profound modification of the formulations and the use of a fused cast process. It was therefore decided to totally revisit the design of these chromium bearing materials, to offer not only a possibility for recycling spent refractories, but also to improve the performance without any compromise on the affordability of our refractory solution. After several years of active research and development, a new and innovative range of chromium bearing materials[5] was developed and is now being industrialized. Because these new materials were specifically designed for the insulation glass-wool furnaces, we chose to name them WOOL50 and WOOL30.

The development of these two new refractory materials started by the realization that most Cr-Al containing refractories were composed of a plurality of grains having very different crystallographic natures and chemistries. Besides, if these materials have quite different microstructures and different crystallographic analysis, all of them are characterized by a quite heterogeneous distribution of chromium oxide throughout the microstructure (cf. figure 4).

---

[4] Cf. Patent US 5,106,795 : Chromic oxide refractories with improved thermal shock resistance, filed by CORHART Refractories (USA).
[5] Cf. Patent application WO2010020344 : Chromium oxide refractory material and patent application WO2012020345 : Chromic oxide powder, both filed by Saint-Gobain CREE (France).

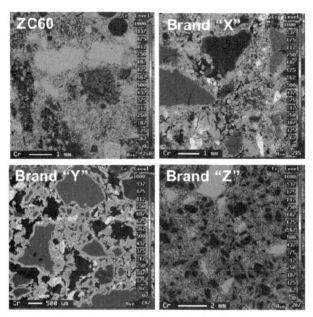

Figure 4. Microprobe elemental maps showing Cr distribution in current commercial refractories.

As previously stated, Zirchrom products are composed of a mixture of fused AZS grains, fused AZS-Cr grains, and dense chrome grains embedded into a corundum-eskolaite matrix of fines. Likewise, products from the SERV, RK "S", or CR ranges are composed of a mixture of sintered or fused corundum, mullite, zircon or zirconia, corundum-eskolaite[6] solid solutions, embedded in a corundum-eskolaite matrix of fines.

It is well known that these individual grains have very different properties and corrosion resistance. For instance. fused AZS grains are close to two times more corrosion resistant than corundum. On the other hand, a high chromium oxide grain exhibits much lower electrical resistivity. Finally, a mullite or zircon grain has a much lower CTE than corundum or eskolaite. We can therefore argue that all conventional materials dot not have similar performance, due to the nature of the grains, however it is commonly believed that the main parameter for the overall corrosion resistance and performance of these Cr bearing refractories is controlled by the total chrome percentage, not the nature of grains.

Our design approach for our new range of material was very original, in the sense that we decided to distribute the chromium oxide in a more homogeneous manner in all phases and grains present in the material, in order to minimize the properties gap. However, our investigations led us to develop a specific and proprietary set of rules[5] for the distribution of the chrome between the grains and the matrix of fines, so as to maximize the performance of our materials. Having the ability to design our own sintered grains formulation while using reclaimed Cr materials was actually a strong advantage, and allowed us develop new materials with optimized microstructures and chrome contents.

---

[6] Chromium oxide (III) lattice structure.

The resulting microstructure of our two new products WOOL30 and WOOL50 are shown in Figure 5. One can, in particular, notice the much more homogenous distribution of Cr throughout the microstructure, as well as the very dense structure of the grains.

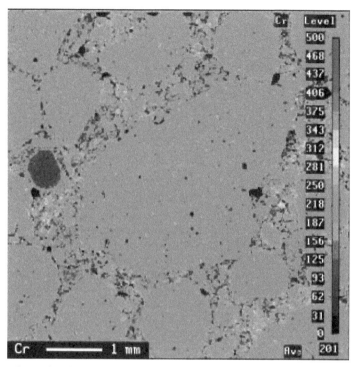

Figure 5. Microprobe elemental maps showing Cr distribution in the WOOL 50.

Although the properties and performances of these two new products will be described great detail in a future communication, we have decided to illustrate our previous statement by showing two very important properties of our new range of materials : corrosion and electrical stability.

As far as corrosion resistance if concerned, we have run several dynamic corrosion tests in an insulation fiber glass at 1450 °C and for 120h, and have reported in Figure 6 the corrosion index (ZC60 being the reference) for the two new materials in comparison with conventional refractories. It can be clearly observed that both new materials exhibit the highest corrosion resistance in both the 30% and 50% chromium oxide range. We firmly believe that this exceptional result is attributed to the original and proprietary Cr distribution pattern that we have used. The analysis of this chart also shows that not all materials in a common Cr content range are equivalent and that a difference of up to 30% can be observed.

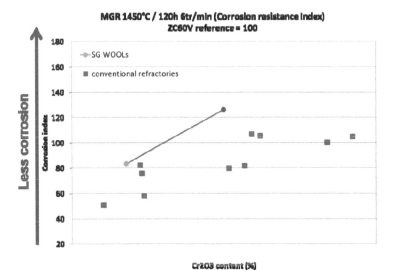

Figure 6. Corrosion resistance in insulation fiber-glass at 1450°C (120h, 6 rpm). (Corrosion index = corroded volume of ZC60 / corroded volume) for the WOOL materials in comparison with conventional refractories. Connected line for SG WOOLS is for general guidance only.

Another illustration of the benefit of the original design of our new materials is the control of the electrical resistivity. The electrical resistivity of Cr bearing materials for insulation glass wool is a very relevant selection parameter when designing an electric melter. A high electrical resistivity should indeed be favored to minimize overheating of the refractories located next to the electrodes due to Joule effect. Overheating can result in, at minimum, excessive corrosion and can put the furnace life at risk in case of electrical run away. To reduce the risk, one may also look at the capacity of the material to be cooled, in order to minimize the effect of the Joule effect. Therefore, thermal conductivity is a very important characteristic for the selection of a refractory. Actually, the combination of these two parameters might be preferred, since we can easily estimate the temperature increase of a refractory located between two electrodes by the following simplified relation:

$$\Delta T \cong \frac{U^2}{ER.\lambda}$$

where ΔT is the temperature increase due to Joule effect, U is the voltage, ER is the electrical resistivity and λ is the thermal conductivity of the refractory at the working temperature.

From this relation it is easily understood that the best material for reducing the overheating is the one with the highest ER x λ. Finally, for the purpose of the selection of materials for electrode blocks and/or the lining of electrical melters, one should compare the performance of the materials both in terms of corrosion resistance of the refractory and ER x λ.

Figure 7 shows the comparative performance of the new WOOL materials in comparison with conventional refractories. This chart clearly shows that the two new materials are best in class in terms of compromise between corrosion resistance and electrical stability, and should be perfectly suited for electric melters, which we believe will soon become the main furnace design for the manufacturing of wool-glass.

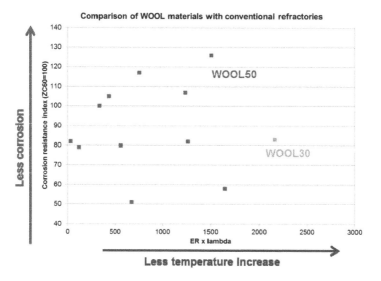

Figure 7. Corrosion resistance vs. ER x λ of the new material compared to conventional refractories.

CONCLUSION

Starting in 2014, SEFPRO will be proposing a full recycling program for spent Cr bearing refractories from insulation glass-wool furnaces. Contrary to other programs available on the market, this new recycling program allows 100% recycling of the spent Cr refractories, provided dismantling is properly performed (glass removal / brick sorting). This program is available for refractories used for the manufacturing of insulation wool glass and containing more than 15% Cr_2O_3 (competitors materials included). It can be offered both in Europe and USA (limited at this time to countries within the OCDE or having ratified the Basel convention). The whole transportation and recycling procedure has been reviewed and approved by the EPA in the US and the INERIS in France. We firmly believe that our solution sets a new standard of environment care and sustainability in the glass refractory industry, due to the following facts:

- Eliminates the need for landfill disposal of dangerous Cr containing wastes

- Reduces the impact of the manufacturing of refractories on the environment (less mining and processing of chrome ore)

- Allows direct recycling in refractories (by using our specific modus operandi for clean and safe dismantling & transportation of spent refractory linings)

- Our proprietary recycling process offers complete hexavalent chrome elimination,.

Simultaneously, a new range of Cr bearing refractories will be introduced to the insulation glass-wool industry. WOOL 30 and WOOL 50 materials not only use reclaimed raw materials, but they also offer an original alternative to conventional refractories by displaying, among other properties, best in class corrosion resistance and electrical stability.

As of today, pre-industrial production started in our Savoie Refractaires facility in Vénissieux (France), several furnaces have been either partially or totally dismantled, and we have produced and delivered several prototypes of WOOL 30 and WOOL50 for qualification purposes in Europe as well as in the US.

This recycling program shows SEFPRO's very strong commitment to the glass-wool industry with several years of Research and Development and multiple major investments under-going in our plants in Europe and in the USA. We expect full commissioning of this equipment by the end of 2014.

FURNACE REPAIR AFTER A HURRICANE FLOODING AT MONTERREY, MEXICO

Roberto Cabrera

Vitro, S.A. de C. V.

ABSTRACT

The recovery of a glass plant of two float lines after hurricane Alex that hit on July 1st, 2010 is presented, furnaces damages are described and how the problems were addressed and eventually overcome the situation to get back to "normal" operation. Bubble glass quality issues presented during and after the recovery of the glass plant are presented.

INTRODUCTION

On July 1st, 2010, Alex, a category 2 hurricane with wind speed of 165 miles/hr, entered the coast of Gulf of Mexico containing lots of rain, as shown in Figure 1. Alex is considered the largest hurricane that has reached Monterrey area ever, with more rain that hurricane Gilberto in 1988. On July 2nd, the normally "Dry" Santa Catarina River (Figure 2) carried out water all the way across its 500 m width.

Figure 1. Hurricane Alex path towards Monterrey area

Figure 2. Santa Catarina river

Flooding Event

VF-1 & VF-2 are two float lines located side by side in Vitro float glass plant at García, NL, Mexico, dedicated to both automotive and architectural glasses.

When constantly raining during the arrival of Hurricane Alex to Monterrey surroundings, the river next to the plant road became dammed-up when floating debris clogged the river's entrance to a bridge, The accumulating water then flooded the streets and flow down the ramp into the basements of both glass furnaces trough the ramps. The ground was already saturated water.

5 meters of water and mud entered into the basement in less than 30 minutes (as shown in Figure 3) creating a great amount of steam that went into the melter through the regenerators and suddenly increasing the furnace internal pressure from 0.25 mm of water column to more than 10 mm.

Figure 3. Water path from river to furnaces basement

Combustion air fans, diesel motors, electric boost transformer and regulator, and other equipment were in the basement. They were covered with water and mud with no possibility to operate. Water at the basement entered the regenerators and tunnel, and blocked the exit of the flue gasses to the chimney. Several days afterwards the water level (Figure 4) was still too high to resume operations.

With the lack of natural gas, both furnaces started to cool down and the main task at that time was to maintain them hot enough to prevent any refractory failure.

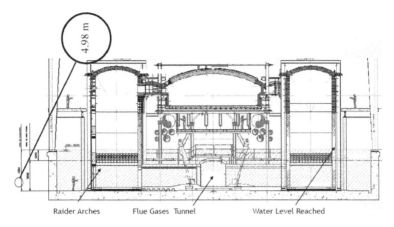

Figure 4. Water level at basement

Furnaces Damages

VF1 & VF2 Back wall silica toes fell down on the glass and froze in place due to lack of enough natural gas. At VF2, due to the cooling down of the waist area, two previously cracked pieces of the waist entrance arch slipped down and fell into the glass.

10 days were required to pump out all the water and clean up the mud from the basement, bottom regenerators and tunnels.

Regenerators and rider arches during cleaning of mud inside after the water level receded. Water literally "washed out" all the sulfates from raider arches

VF1 suspended wall (shown in Figure 5) was the most damaged part. Even after steel structure reinforcement, the refractory was too damaged to continue operating. The decision was to stop the furnace for a cold repair. A Merkle suspended wall steel structure was bought from a float furnace stopped in the USA. Merkle supplied the refractory material within a couple of weeks. Some adjustments had to be done to the length of doghouse to allow the charger to operate properly. Hotwork was the first supplier to arrive at the plant, and the last to leave few months later. VF1 started operation on Sep 11, 2010, after almost two and a half months since the flood. By Sep 20th, it was already producing automotive quality glass.

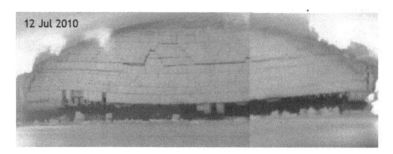

Figure 5. VF1 suspended wall

VF2 suspended wall was less damage than VF1. Air cooling was re-established in 3-4 days after the flood and steel structure was also reinforced with an additional dead beam on top of the existing one. Silica toes fell down in the same way as happened at VF. The wall was bent in the middle and, in order to stop radiation towards the steel structure, the toes had to be installed again. 27 silica toes installed at the right hand side of the wall. All silica toes were installed and missing silica blocks located in position. Silica ceramic welding was applied to all the wall joints to prevent air from leaking into the melter. VF2 Furnace started up on July 23rd.

VF2 Bubble Outbreak

By this time a significant amount of bubble defects were present in the glass ribbon. With the waist stirrers off, bubbles were present mostly in the right half of the ribbon.

During an inspection at furnace waist: Foam was accumulated on the RHS waist pipe. When removed, the pipes presented signs of accumulated bubbles in a very specific position. A very defined line of foam was seen at the working- end on the RHS. Bubble tracking by mathematical modeling was performed to identify the possible location of the source. Everything pointed out to the missing blocks of the waist arch (shown in Figure 6), the bubble gas analysis was consistent with a contamination of Silicon and Aluminum, coming from the unreacted ceramic weld material previously applied to this arch.

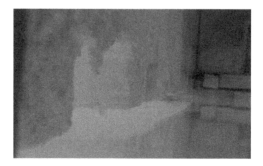

Figure 6. Two Pieces of refractory blocks missing from furnace waist arch

In order to try to take the pieces out, the glass was drained to the lowest possible level. Once reached the minimum level, it was possible to see the two ~90 kg blocks sitting on the bottom of the furnace waist.

Several tools and devices were manufactured and tested to take out the pieces from the furnace. Finally on Saturday, August 14th, it was possible to take the blocks out from the waist, after several days of failed trials. Temperature at this area was around 1200°C.

When the block was inspected, it was found that ceramic weld material was attached to the block as thick as 4 to 5 inches. A large amount of bubbles were frozen off when the glass cooled down. It was possible to see the bubble generation coming out from this material, as shown in Figure 7.

Figure 7. Bubbles generation from ceramic weld material

VF2 Operation re-started on Aug 22nd, a month later from the previous start up, and almost two months after the flood. Bubble defects levels went down; however, the level was still high for some glass qualities. Chemical gas bubble analysis showed that there must be some remaining contamination material still located in the right side of the waist area.

In order to inhibit the mechanism of bubble generation of the remaining ceramic weld material, a waist cooler was proposed to reduce glass temperature and increase the viscosity. After the coolers installation the fault index went down from values of 4.0 – 5.0 to 1.0 – 1.2 Defects/10m^2. By October, VF2 furnace was producing close to the usual 95% yield, as demonstrated in Figure 8.

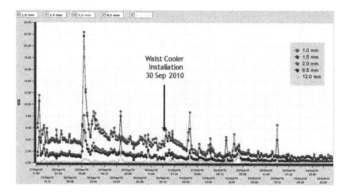

Figure 8. Bubble defects recovery after water cooler installation

SUMMARY

Key Factors to Overcome the Situation:

- Management Support. *Trust and confidence of Top Management* over the road, and tough decision making was always present.

- A Group of Suppliers. We count the many companies that are just there when you need them. *Hotwork, Fosbel, Glass Service, Merkle, Glass Design, Mirsa among others,* are examples of the type of companies that helped us during that time. W*ithout them the case would be a total failure.*

- Technical Consultants. Many meetings and discussions were held with consultants to *think things with different perspective*.

- Innovation and Creativity. New ways to deal with problems, *unproven solutions were implemented and modern techniques used*, such as mathematical modeling applied to find out the source of bubbles, practical bubble tracking, new tools developed and tested.

- Perseverance – Never give up.

- People. At the end, *It is always about People* to overcome every single problem and catastrophe.

It took several months for the Vitro team to sort out the total condition of the plant. Many extra hours per day and weekend work time were required but we continued producing glass to serve our customers, which is the very bottom line.

Author Index